T0273334

Aquatic Environmental Systems

Considering that environmental science draws students and practitioners with widely varied backgrounds, there is a need for materials that help readers to grow their knowledge of fundamental principles from chemistry, physics, and biology to understand, describe, and predict the ways in which constituents (sediment, nutrients, organic matter, etc.) interact and move in aquatic systems (rivers, lakes, groundwater, and the atmosphere). *Aquatic Environmental Systems: An Interdisciplinary Approach for Scientists and Engineers* focuses on developing a common vocabulary and a rigorous material balance-based approach to understand these movements and interactions. It examines the key properties of water and the ways they impact the behavior of water in the environment, providing a focused enumeration of those aspects of water structure that have direct and profound impacts on aquatic environmental systems.

FEATURES:

- Provides open-ended examples to allow students to tailor work to their personal local/regional interests.
- Focuses on conveying understanding of the underlying principles and assumptions/limitations which are frequently underemphasized or overlooked entirely in other books.
- Deemphasizes straight memorization while focusing on methods that can be applied to more broad-based problem solving.
- Accommodates a wide range of mathematics skills and backgrounds.

Aquatic Environmental Systems

An Interdisciplinary Approach
for Scientists and Engineers

Roger C. Viadero, Jr.

CRC Press
Taylor & Francis Group
Boca Raton London New York

CRC Press is an imprint of the
Taylor & Francis Group, an **informa** business

Designed cover image: Shutterstock

First edition published 2024
by CRC Press
2385 NW Executive Center Drive, Suite 320, Boca Raton FL 33431

and by CRC Press
4 Park Square, Milton Park, Abingdon, Oxon, OX14 4RN

CRC Press is an imprint of Taylor & Francis Group, LLC

ISBN: 978-1-032-26718-0 (hbk)
ISBN: 978-1-032-26721-0 (pbk)
ISBN: 978-1-003-28963-0 (ebk)

DOI: 10.1201/9781003289630

Typeset in Times
by Apex CoVantage, LLC

*To my wife, Patty, and my children, Avery, Peyton, and Spike.
Now that I've finished writing this book, I promise to [try to] stop
spontaneously talking about hydrogen bonding, alluvial fill, and the
great value of clean water while you're trapped in my truck on road
trips. I cannot, however, vow that I won't occasionally stop at random
places to look at an interesting stream, wetland, or other body of water.*

Contents

About the Author

Roger C. Viadero, Jr. is Professor and Director of Western Illinois University's (WIU) Institute for Environmental Studies (IES) where he chairs the interdisciplinary Ph.D. program in environmental science. Dr. Viadero is an aquatic environmental engineer who specializes in the remediation of environmental systems that have been adversely impacted by human activity. Given the broad scope of his work, Roger typically works with a multidisciplinary team of biologists, engineers, geographers, geologists, and planners, among others. Prior to joining WIU, Viadero was Robert C. Byrd Associate Professor in the Department of Civil and Environmental Engineering at West Virginia University, where he directed the university-wide Center for Environmental Research.

Roger has served as President of the Aquacultural Engineering Society and is a member of the editorial board of the *Journal of Aquacultural Engineering*. He has also worked as a science adviser and panel manager for the Aquaculture and Aquatic Products Small Business Innovation Research (SBIR) Program through the US Department of Agriculture's National Institute for Food and Agriculture. Professor Viadero is a board-certified environmental engineering member (BCEEM) of the American Academy of Environmental Engineers and Scientists with a specialization in hazardous waste management and site remediation and is an Ecological Society of America-certified senior ecologist.

Acknowledgment

I am especially grateful to Professor Ronald Fortney for his friendship, mentorship, and willingness to help an engineer broaden his botanical horizons beyond "Christmas tree" and "not a Christmas tree." Every time I pull on my waders and head out to the field with my students, I am reminded of his limitless enthusiasm for our work, his dedication to meaningful interdisciplinary team building, and the sacrifices he made to mentor so many of us into the scientists, engineers, teachers, and people we are today.

Introduction

As a broadly trained aquatic environmental engineer, I've specialized in the remediation of natural systems that have been impacted by human activity. This work has necessarily bridged the gaps and overlaps between and among the fields of environmental engineering, environmental science, and ecology. Consequently, most of my professional experience has involved multidisciplinary teams of biologists, engineers, geographers, geologists, and planners, among others. This has left me with the strong belief that our collective ability to advance the state of science and, ultimately, the quality of aquatic and adjacent environmental systems will come down to our ability to work at the interface of many academic disciplines. While this might sound straightforward, there is still a need for resources that integrate these broad areas of study while still maintaining enough focus to be useful to a multidisciplinary audience. This was my goal for writing this book.

We'll begin by considering basic concepts from ecology and developing a working definition of aquatic environmental systems. This will be followed by a consideration of the unique properties of water and how they impact its behavior in the environment. Then, we'll look into the ways water moves through the environment and acquires and imparts characteristics from and to the surrounding environment. We'll broaden our consideration to include the ways in which water shapes the earth with a particular focus on alluvial channels. Once we establish a basic understanding of water properties, sources, distribution, cycling, and hydrology/fluid mechanics, we'll spend time considering the physical, chemical, and biological characteristics of natural waters followed by chemical reactions and reaction mechanisms. Building on this information, we'll consider reaction kinetics and models of ideal reactor systems. From this point, we'll be able to quantitatively describe the movement of non-reacting and reacting constituents in static and flowing water systems. After developing idealized models, we'll broaden our consideration to examine and account for the ways observed behavior in natural systems can deviate from idealized predictions.

While natural bodies of fresh water will be the "continuous phase" in the vast majority of our analyses, terrestrial and atmospheric environmental compartments certainly won't be neglected. Rather, these compartments will be considered to the extent that they impact and/or are impacted by aquatic systems—typically at land-water and air-water interfaces.

Whenever possible, I have presented both the discrete and differential approaches to the required mathematics. This is deliberate. While there are a number of problem-solving techniques that require a strong working knowledge of calculus and differential equations, there's no reason for everyone else to be left in the dark. Likewise, I've worked examples in more detail than is shown in many other textbooks. This too is deliberate. Since there's been a huge shift in the way we teach and learn, I think it's important to take some of the mystery out of how we go from

DOI: 10.1201/9781003289630-1

Equation X to Equation Y. Likewise, I've tried to integrate a range of assumptions into the examples. As a student, I recall spending a lot of time wondering why a certain assumption was made under one set of circumstances but not under another seemingly similar situation. I hope this material serves as a starting point as you work to identify and apply often nuanced assumptions to more complicated aquatic environmental systems.

1 An Introduction to Freshwater Environmental Systems

1.1 FRESHWATER ENVIRONMENTAL SYSTEMS ARE IN EVERYONE'S WHEELHOUSE!

The topic of aquatic environmental systems isn't the sole domain of any one academic or professional discipline. For example, the study of a single stream reach is likely to involve aquatic ecologists; environmental scientists; civil, environmental, and water resource engineers; geologists; geomorphologists; and fisheries biologists, among other professionals. The breadth of this field is what makes this discipline both exciting and complicated to study.

We're going to focus largely on the rivers and streams that serve as the Earth's circulatory system. This also includes consideration of the interface between aquatic and terrestrial habitats in addition to other types of waterbodies such as lakes, ponds, and wetlands. Since this is such an interdisciplinary field, there is almost a limitless number of ways we might organize our study freshwater environmental systems. Should we focus on fluid mechanics, water chemistry, aquatic ecology, or fluvial geomorphology? In fact, we actually need to have a working knowledge of these fields and others. For example, being competent in the use of geographic information systems software might seem like it's tangentially related to our field of study; however, this is a valuable skill that can be especially useful when trying to understand the ways different phenomena are related on the landscape over time.

While the exact definition varies, most people agree that a system is a set of parts that work together according to a set of rules to achieve a common goal. According to Odum (1964), an ecosystem is the basic functional unit of organisms and their environment. This includes the way organisms interact among themselves and with all other biotic and abiotic constituents. In this respect, scale and size are less important than the interactions between biotic and abiotic parts of the system. So, an ecosystem can range in size from an entire watershed to a pond or a small unit volume of a stream.

1.2 ECOSYSTEM STRUCTURE

The structure of an ecosystem refers to the abiotic conditions/factors and the living organisms that make up an environmental system. Of interest are the relationships between living organisms and the physicochemical environment. For example, fish,

DOI: 10.1201/9781003289630-2

invertebrates, and hydrophytic vegetation interact with a range of water temperatures, clarities, and nutrient levels, all comprising parts of the structure of a freshwater aquatic system.

1.2.1 ABIOTIC FACTORS

Abiotic factors are all of the non-living parts of an environmental system. These include temperature, radiant energy from the sun, humidity, wind, water, and chemical composition. The total of all abiotic factors in a system is known as the standing load. Abiotic factors create boundary conditions that act as limits on parts of the ecosystem. For example, high wind speeds and low temperatures can lead to stunted plant growth. Likewise, cacti and succulents are unlikely to thrive in areas that receive an abundance of precipitation. Similarly, some plants such as rhododendron and holly thrive in acidic soils. The concept of a boundary condition will become important as we work to develop approaches to quantitatively describe interactions in aquatic environmental systems.

1.2.2 BIOTIC COMPONENTS

Biotic components include all living organisms present in an environmental system. These are broadly classified based on their ability to produce their own food. Autotrophic organisms are able to create their own food from sunlight or inorganic chemicals. Photosynthetic autotrophs contain chlorophyll (*or an equivalent pigment*) that allows them to absorb photosynthetically active radiation (light with a wavelength ranging from 400 to 700 nm) from the sun. The energy from the sun is used to metabolize water and carbon dioxide to produce oxygen gas. Examples of photosynthetic autotrophs in aquatic systems include green algae (*Chlorophyta* and *Charophyta*), cyanobacteria, and phytoplankton.

Chemosynthetic autotrophs obtain their food from inorganic compounds. For example, the aerobic oxidation of ammonia to nitrate is facilitated by two chemoautotrophic bacteria (*Nitrosomonas* and *Nitrobacter*). *Nitrosomonas* utilize oxygen to oxidize ammonia to nitrite. Additional oxygen is then used by *Nitrobacter* to oxidize nitrite to nitrate. Through this process, plants receive the nitrate needed to form amino acids and subsequently create the proteins needed for cell growth. In contrast, heterotrophic organisms derive their energy and nutrients by consuming plants and animals.

1.2.3 COMPARTMENTS

Structure also involves the various compartments in an environmental system; these include the atmosphere, the biosphere, the hydrosphere, the pedosphere, and the lithosphere. The atmosphere (air) is the gaseous layer that envelops the Earth. The atmosphere is made up of nitrogen (78.1%), oxygen (21.0%), and argon (0.9%) gases in addition to small fractions of carbon dioxide, hydrogen, helium, and other gases. In addition to gases, the atmosphere contains water vapor as well as small particles.

The biosphere includes every place on Earth where life exists. This includes locations that are home to some surprisingly rugged life forms. For example, *Thiobacillus ferrooxidans*, *Ferrobacillus ferrooxidans*, and *Thiobacillus thiooxidans* help to catalyze the oxidation of ferrous iron at low pH (Ashmead 1955; Leathen and Bradley 1954; Gleen 1950). In addition to being acidophilic, these bacteria are microaerophiles and require very little oxygen to initiate the oxidation of iron in pyrite ($FeS_2(s)$) (Snoeyink and Jenkins 1991).

The hydrosphere (water) encompasses every location on Earth that contains water. Since ~71% of the Earth's surface is covered by water, the hydrosphere plays a key role in the movement of matter and energy in the environment (Shiklomanov 1993). The pedosphere (soil) is the part of the Earth's crust that contains soil and exhibits active soil-forming processes. The lithosphere (rocks) is the solid outer section of Earth and the upper mantle. It extends from the Earth's surface to a depth of about 97 km (~60 mi). The lithosphere is the most rigid part of the Earth. In some cases, the pedosphere and lithosphere are treated as one compartment. However, since soils are dynamic entities that play an important role in the movement, storage, and interaction of water, there is a substantial case for maintaining a distinction between the pedosphere and the lithosphere.

Each sphere has its own distinctive characteristics and boundaries. However, in the environment, these boundaries often overlap. In the context of aquatic environmental systems, the relationships among these compartments are most easily seen by considering the hydrologic cycle, which is discussed in detail in Chapter 3.

1.3   ECOSYSTEM FUNCTION

The definition of ecosystem function is open to a little more interpretation than that of ecosystem structure. For example, some consider ecosystem function to describe the physical, chemical, and biological processes that transform and move matter and energy inside and between environmental compartments (Naeem 1998). A few examples of ecological function in an aquatic environmental system include the movement of solid and soluble constituents in water, continuously reshaping the pattern and profile of alluvial channels, and the cycling of nitrogen, phosphorus, carbon, and other elements such as sulfur.

Others regard ecosystem function as the net ability of the parts and processes in an ecosystem to provide goods and services that satisfy a human need (de Groot *et al.* 2002). Clearly, the term "goods and services that satisfy human need" provides ample room for interpretation. The National Research Council offers a reference that provides detailed coverage of this approach (NRC 2005).

1.4   A HIERARCHY OF FUNCTIONS IN AQUATIC ENVIRONMENTAL SYSTEMS

One useful approach to studying aquatic environmental systems is a hierarchical framework that's been used to set objectives for stream restoration work in the US (Figure 1.1) (Harman *et al.* 2012; Fischenich 2006). While the focus of this approach is on steam restoration, the framework is a useful way to study how ecological

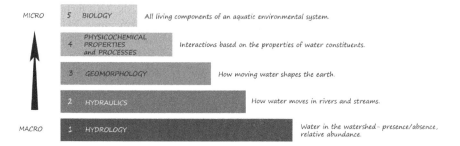

FIGURE 1.1 Functional relationship between structural and functional elements of river and stream ecology (after Harman *et al.* 2012; Fischenich 2006).

functions facilitate the movement of matter and energy within and between the structural elements of aquatic environmental systems.

Using this approach, hydrology—the movement of water in a watershed to streams—forms the base of the stairway. This functional category allows us to describe the origins and fate of water. In the absence of suitable hydrology, we wouldn't have aquatic environmental systems. Once water makes its way into streams and rivers, the hydraulics or the movement of water in channels and through stream sediments is the next level in this functional hierarchy. This provides a means for the subsequent shaping of landforms and movement of matter by/with/in water (geomorphology). The final major factors that support the biological composition of an aquatic environmental system are the physicochemical properties and processes that occur in streams. These include water and air temperatures, the availability of dissolved oxygen, the cycling of nutrients, the processing of organic matter, *etc.*

Historically, many environmental assessment methods focused on specific biological indicators without considering the impacts of/for underlying processes and functions (Somerville 2010). In this regard, it is notable that biology sits at the top of our stairway. This reflects the dependence of an assemblage of organisms on a unique combination of factors that are characterized in the first four levels. For example, during protracted droughts, a change in hydrology will necessarily impact hydraulics, which will affect geomorphology.

Example: The Impact of Drought on Organisms in the Colorado River Basin—from Hydrology to Biota

Since 2000, the Colorado River Basin has been in a 20+-year-long drought cycle that is exacerbated by the effects of global climate change. Scientists believe this is the most significant drought in the region in almost 1,200 years. Clearly, a drought will have direct adverse impacts on the abundance and availability of both surface and ground water. In the Upper Colorado River Basin (Wyoming, Colorado, and Utah), it's estimated that 56% of streamflow comes from groundwater discharge (Bruce *et al.* 2015; Miller *et al.* 2014). The reduced abundance of water in the watershed corresponds to a decrease of in-channel flows, which in turn affects in-stream hydraulic processes, including the movement of bed and bank material.

In some cases, streams that previously existed year-round might shift to a more ephemeral phenological pattern. In other cases, the total area of water habitat can decrease as water is depleted faster than it can be recharged. These changes also have direct impacts on many physicochemical characteristics of water and related processes. For example, when water is lost in a stream, aquatic organisms can be exposed to higher levels of ambient radiation as well as elevated water temperatures (Friggens *et al.* 2013). In the Colorado River Basin, the collective impact of these changes has limited the extent and resulted in the fragmentation of populations of native cutthroat trout (Ma *et al.* 2023).

In practice, this approach is often used to decide which supporting functions need to be addressed in order to restore a particular function to a stream. For example, if a goal of stream restoration is to better describe and manage the overland and subsurface flow of water to a channel, lake, or wetland, efforts should be focused on Level 1—Hydrology. If a goal is to increase the biomass of a particularly sensitive species, a Level 5 concern, a much more broad-reaching assessment and consideration must be made of hydrologic, hydraulic, geomorphological, and physicochemical functions of the system. Clearly, Levels 2 through 5 in Figure 1.1 each depend on factors that control the lower steps. This is intentional, as it represents the connections and dependence of each factor on other structural elements and functions of the system. This is simply the way things work in the natural world.

REFERENCES

Ashmead, D. (1955). "The Influence of Bacteria in the Formation of Acid Mine Waters," *Colliery Guardian*, 694–698.

Bruce, B., Clow, D., Maupin, M., Miller, M., Senay, G., Sexstone, G., and D. Susong (2015). "US Geological Survey National Water Census—Colorado River Basin Geographic Focus Area Study: US Geological Survey Fact Sheet 2015–3080," 4 p. http://dx.doi.org/10.3133/fs20153080.

de Groot, R., Wilson, M., and R. Boumans (2002). "A Typology for the Classification, Description and Valuation of Ecosystem Functions, Goods and Services," *Ecological Economics*, 41, 393–408.

Fischenich, J. (2006). *Functional Objectives for Stream Restoration, EMRRP Technical Notes Collection (ERDC TN-EMRRP-SR-52)*, US Army Engineer Research and Development Center, Vicksburg, MS.

Friggens, M., Finch, D., Bagne, K., Coe, S., and D. Hawksworth (2013). "Vulnerability of Species to Climate Change in the Southwest: Terrestrial Species of the Middle Rio Grande, Forest Service General Technical Report, RMRS-GTR-306," U.S. Department of Agriculture, Forest Service, Rocky Mountain Research Station, Fort Collins, CO.

Gleen, H. (1950). "Biological Oxidation of Iron in Soil," *Nature*, 166, 871–873.

Harman, W., Starr, R., Carter, M., Tweedy, K., Clemmons, M., Suggs, K., and C. Miller (2012). *A Function-Based Framework for Stream Assessment and Restoration Projects*, US Environmental Protection Agency, Office of Wetlands, Oceans, and Watersheds, Washington, DC, EPA 843-K-12–006.

Leathen, W., and S. Bradley (1954). "A New Iron-Oxidizing Bacterium: *Ferrobacillus Ferrooxidans*," *Bacteriological Proceedings*, 44.

Ma, C., Morrison, R., White, D., Roberts, J., and Y. Kanno. (2023). "Climate Change Impacts on Native Cutthroat Trout Habitat in Colorado Streams," *River Research and Applications*, https://doi.org/10.1002/rra.4122.

Miller, M., Susong, D., Shope, C., Heilweil, V., and B. Stolp (2014). "Continuous Estimation of Baseflow in Snow Melt Dominated Streams and Rivers in the Upper Colorado River Basin: A Chemical Hydrograph Separation Approach," *Water Resources. Research*, 50, 6986–6999.

Naeem, S. (1998). "Species Redundancy and Ecosystem Reliability," *Conservation Biology*, 12, 39–45.

National Research Council (2005). *Valuing Ecosystem Services: Toward Better Environmental Decision-Making*, The National Academies Press, Washington, DC, 290 pp., https://doi.org/10.17226/11139.

Odum, E. (1964). "The New Ecology," *BioScience*, 14, 14–16.

Shiklomanov, I. (1993). "World Fresh Water Resources," in Gleick, P. (Ed.), *Water in Crisis: A Guide to the World's Fresh Water Resources*, Oxford University Press, New York.

Snoeyink, V., and D. Jenkins (1991). *Water Chemistry*, John Wiley and Sons, New York, 480 pp.

2 Water
Properties and Behavior

2.1 THE MOLECULAR STRUCTURE AND BONDING OF WATER MOLECULES

Water—H_2O—is formed when two hydrogen atoms are covalently bonded to one oxygen atom, as shown in Figure 2.1. This leaves four unpaired electrons in the p-orbital of the oxygen atom. You will recall that covalent bonds form when atoms share electrons. Among the different types of chemical bonds, covalent bonds are the strongest. Consequently, a relatively large amount of energy is needed to break the covalent bonds in a water molecule. For example, the complete dissociation of water into its constituent ions is a two-step process. First, one proton is separated from the water molecule, requiring 493 kJ/mol to break the HO-H bond. This process is followed by breaking the O-H bond, which requires an additional 424 kJ/mol.

While water molecules have zero net charge, the four unpaired electrons in the p-orbital of the oxygen atom result in a region of locally high negative charge (2δ-) in the vicinity of the oxygen atom. Similarly, by sharing an electron with the oxygen atom, the net charge around each of the two hydrogen atoms becomes less electronegative, which results in a net positive charge (δ+), as shown in Figure 2.2. The formation of these regions of local charge concentration makes water a polar compound.

In three dimensions, water is a tetrahedrally shaped molecule. To visualize this, you need to include the two sets of unpaired electrons, as presented in Figure 2.3. For a regular tetrahedron, the central angle is $109.5°$. However, the asymmetric distribution of charge in a water molecule results in a central angle of $104.5°$, which is slightly less than expected for an ideal tetrahedron. When multiple water molecules are present, they orient themselves so that the regions of partial positive charge are aligned with regions of partial negative charge in adjacent water molecules.

Hydrogen bonds form when a hydrogen atom that's covalently bonded to an electronegative atom such as oxygen is attracted to the electrons on an atom in a neighboring molecule. In liquid water, hydrogen bonds are constantly formed and broken as water molecules move. In fact, the lifetime of a single hydrogen bond in water is reported to be on the order of picoseconds (10^{-12} s). While the energy required to break a hydrogen bond between an oxygen and a hydrogen atom is only 21 kJ/mol, hydrogen bonding plays a more significant role in the behavior of liquid water than we would predict based on its strength alone. In the case of water, a net negative charge equal to that of four electrons allows each water molecule to form four hydrogen bonds with nearby water molecules.

The ability to form, break, and reform hydrogen bonds between molecules gives water a number of its unique properties. For example, water's ability to continuously form and break hydrogen bonds allows it to absorb large amounts of

DOI: 10.1201/9781003289630-3

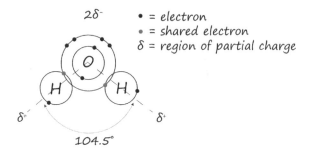

FIGURE 2.1   Orbital diagram of a water molecule.

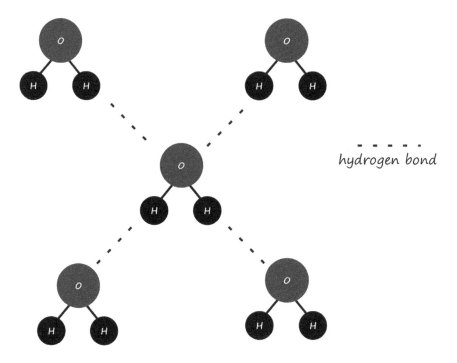

FIGURE 2.2   One water molecule hydrogen bonded to four additional water molecules.

energy—especially when compared with other materials such as metals. This property is reflected by the specific heat capacity, C_s, or the amount of heat required to raise the temperature of one gram of a substance by one degree Celsius. The C_s values for water in its three phases as well as a few other materials are presented in Table 2.1.

Consider a bucket of water and a steel baking sheet that are placed next to one another and left in the hot summer sun. Would it be possible to hard boil an egg in the bucket of water? Could an egg be fried on the baking sheet? Intuitively, we all know that the water would never get hot enough to boil. Kids go to the community pool to

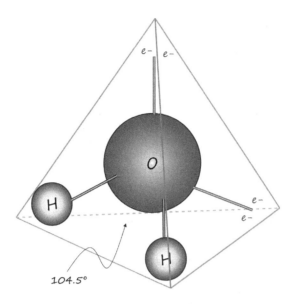

FIGURE 2.3   Tetrahedral shape of a water molecule.

TABLE 2.1
Specific Heat Capacities of Freshwater and
Other Materials (Haynes 2014; NIST 2020).

Substance	C_s, J/g°C
Ice (0 °C)	2.09
Liquid water	4.184
Steam (100 °C)	2.03
Aluminum	0.921
Copper	0.377
Dry soil	0.800
Wet soil	1.450
Wet mud	2.512
Sand	0.830
Steel	0.502

---

cool off in the summer—not to be boiled! Water has a huge capacity to absorb heat energy. It also releases heat energy slowly. In contrast, steel has a specific heat capacity that's 92% lower than that of water. Depending on the relative humidity and other atmospheric factors, the steel sheet can reach a temperature of over ~93 °C (200 °F). Since a temperature of ~71 °C (160 °F) is recommended to fry an egg (Egg Safety Center 2020), you'll need to keep a close eye on your improvised stovetop!

Example: The Energy in Water

How much energy is needed to increase the temperature of water in an Olympic-sized pool from (24 °C) (~75 °F) to its boiling point? Let's begin by listing the things we know—or can find. A regulation Olympic swimming pool is 50 m long, 25 m wide, and 2 m deep. This corresponds to a volume of 2,500 m³ or 2.5 million liters. We want to increase the temperature of liquid water by 76 °C (from 24 to 100 °C). The temperature range that was specified does not involve a change in the phase of water, so our calculations can be based on the specific heat capacity of liquid water. Based on this information, we can use dimensional analysis (Equation 2.1) to find the amount of energy needed to increase the temperature of 2,500 m³ of water by 76 °C:

$$\text{Equation 2.1} \quad 2500\,m^3 \left(1\frac{kg}{m^3}\right)\left(4.184\frac{J}{g\,°C}\right)\left(10^3\frac{g}{kg}\right)(76\,°C) = 7.95x10^{11}J \cong 8x10^{11}J$$

Is this a large or small amount of energy? To put this into perspective, how long would it take for a 1 MW power plant to provide the necessary energy? A 1 MW power plant produces 1 million Joules of energy each second. Since we know the amount of energy needed and the rate at which energy is produced, we can find the time required (Equation 2.2).

$$\text{Equation 2.2} \quad \frac{8x10^{11}J}{10^6\frac{J}{s}} = 8x10^5\,s = 222\,hr = 9.25\,d$$

So, it would take a 1 MW power plant over nine days to generate enough energy to increase the temperature of the water in an Olympic-sized pool from 24 °C to its boiling point.

Since the strength of hydrogen bonds is much lower than that of covalent bonds, liquid water molecules can rearrange themselves in response to changes in net energy while maintaining some structure. In fact, the continuous formation and breaking of hydrogen bonds allows liquid water to remain condensed when other molecules of similar molecular weights are gases at room temperature (*e.g.*, ammonia). When liquid water is heated, water molecules absorb additional energy and move faster. As water approaches its boiling point of 100 °C, liquid water molecules absorb sufficient energy to break free and become water vapor. Due to the random nature of molecular motion, some water molecules can actually escape from the liquid phase at temperatures that are lower than the boiling point. The amount of energy needed to break the hydrogen bonds in liquid water is known as the heat of vaporization (40.65 kJ/mol). While in the vapor phase, water molecules move around independently. However, as heat energy is lost, the vapor-phase water molecules slow down and condense, allowing hydrogen bonds to form between liquid water molecules.

As liquid water is frozen into ice, the attractive and repulsive electrostatic forces become balanced and the movement of water molecules slows down even more. Water molecules and their corresponding hydrogen bonds become fixed in a three-dimensional lattice where one water molecule is surrounded by four other water molecules. The system remains fixed until energy is added and the ice melts. Due to the polar nature of the water molecules, there are gaps in the ice lattice. As a result, solid water has a density that is less than liquid water. This is another significant difference from other materials in which the solid phase is generally denser than the liquid phase.

Water is often called a universal solvent. This property of water is derived from the combination of its polar structure and its ability to form and break hydrogen bonds. Given their relatively small size, many water molecules are able to surround charged or polar solutes and form hydrogen bonds. When combined with the electrostatic forces that exist between ions or poles of opposite charges, the solute is liberated and enters the aqueous phase. For example, when table salt (NaCl) is added to water, the monovalent sodium and chloride ions are surrounded by water molecules and are transported into solution.

When a nonionic solute such as sucrose ($C_{12}H_{22}O_{11}$) is added to water, polar hydroxyl groups (R-OH) are able to form hydrogen bonds with water. In this case, the oxygen atom is more electronegative than the proton or the carbon atom. As a result, the electrons in the covalent bonds are concentrated around the oxygen and two polar covalent bonds are formed (O-H and O-C). The significance of this behavior of water is especially important given the presence of hydroxyl groups in carbohydrates, nucleic acids, and some amino acids. In contrast, nonpolar substances such as fats and oils are not soluble in water. In these cases, the thermodynamic affinity between water molecules is more favorable than the interaction of water molecules with nonpolar substances.

2.2   PRACTICAL ASPECTS OF THE BEHAVIOR OF WATER

Now, let's consider how the unique properties of water impact life by regulating the temperature on Earth. According to the World Meteorological Organization (2016), Earth has temperature extremes of −89.2 and 57.7 °C (−129 and 134 °F). For comparison, temperatures on the moon's surface can range from −179 to 116 °C (−290 to 240 °F) (Vasavada *et al.* 2016). Water on the surface of the earth traps heat during the day and releases it slowly at night, which helps to maintain a relatively narrow range of temperatures. In the absence of water, the moon is unable to assimilate the large amount of energy imparted by the sun. Consequently, the surface temperature of the moon increases rapidly when exposed to the sun. Since very little heat is stored, temperatures on the surface of the moon fall abruptly when directly exposed to solar radiation.

Now let's consider the temperature profile of large land masses that have relatively little surface water. According to the US National Aeronautics and Space Administration (NASA), temperatures in desert biomes range from daily highs of around 38 °C to nightly lows of −4 °C. These biomes receive around 25 cm of precipitation on an annual basis. The specific heat capacity of land is around 24% of the C_s for liquid water (Table 2.1). Consequently, the temperature of large land masses changes more rapidly than that of comparably sized waterbodies. From a more practical point of view, anyone who's been to the beach in the summer has noticed that the sand can become very hot while the water remains at a much cooler and relatively constant temperature.

Other unique properties of water drive physical processes that are vital to life on Earth. For example, when ice forms at the surface of waterbodies, it provides an insulating layer that allows many aquatic organisms to survive the winter in deeper water. Take a minute to think about what would happen if ice was denser than liquid water. The density difference between liquid and solid phase water also plays an important role in the physical weathering of geological formations. Liquid water can be found in small cracks and pores in rocks. When the water freezes and expands, the ice acts like a wedge. When the ice melts, the liquid water carries away small fragments of rock.

Based on water's ability to dissolve a wide range of materials, minerals are stripped from rocks and soils by flowing water. Major ions found in waters on a worldwide basis are presented in Table 2.2. While ions such as calcium and magnesium are predominant elements found in the Earth's crust, local geochemical characteristics of geologic formations can have an important impact on water quality. For example, Karst (geological substrates that are easily eroded or solubilized) features are made of limestone ($CaCO_3$), dolomite ($CaMg(CO_3)_2$), and gypsum ($CaSO_4 \cdot 2H_2O$). Over time, landforms in Karst regions develop subsurface voids that include caves and sinkholes. The Upper Mississippi River Basin states of Wisconsin, Minnesota, Missouri, Iowa, and Illinois contain carbonate-bearing rock at/near the surface as well as deeper deposits that lay from 15 to 90 m (~50 to 300 ft) below insoluble sediments that were deposited as the glaciers receded. As a result, natural waters in this region are often rich in calcium and have high/excess [carbonate] alkalinity (the ability of a water to resist a change in pH when exposed to an acid; Doctor and Alexander 2021). In contrast, the coalfields of West Virginia and eastern Kentucky

TABLE 2.2
Major Ions and Common Sources Found in Water throughout the World.

	Ion	Representative Sources
Cations	Calcium, Ca^{2+}	Limestone ($CaCO_3$), gypsum ($CaSO_4$), and dolomite ($CaMg(CO_3)_2$), some clay minerals, and feldspars.*
	Magnesium, Mg^{2+}	Dolomite, magnesite ($MgCO_3$), and some clays.*
	Potassium, K^+	Feldspars, some clay minerals, ocean water, and industrial waste.
	Sodium, Na^+	Feldspars, halite ($NaCl$), some clay minerals, ocean water, and industrial waste.
Anions	Bicarbonate, HCO_3^-	The interaction of carbon dioxide and water on carbonate-containing rock including limestone, dolomite, and magnesite.
	Sulfate, SO_4^{2-}	Dissolved rocks and soil containing gypsum, iron pyrite (FeS_2), and similar sulfur-containing minerals compounds.
	Chloride, Cl^-	Ocean water, brines, halite, sedimentary and igneous rock, and industrial wastes.

* Ca^{2+} and Mg^{2+} are major sources of water hardness. Dissolved iron and aluminum are also sources of water hardness.

contain very little Karst material. The Karst deposits in this region are also generally located closer to the surface (Doctor *et al.* 2015). In addition to Karst minerals, this region is rich in iron pyrite (FeS_2). When exposed to oxygenated water or air, Fe^{2+} in iron pyrite is oxidized to ferric iron, Fe^{3+} which precipitates as $Fe(OH)_3(s)$ at circum-neutral pH values. The presence of ferric iron is easily identified by orange staining of streambeds, as shown in Figure 2.4.

FIGURE 2.4 Staining on rocks and the streambed immediately downstream from Elakala Falls (Blackwater Falls State Park, Davis, WV) due to ferric hydroxide precipitation. Photo by Roger C. Viadero, Jr., June 2, 2022.

In some cases, the natural characteristics of water are reflected in the extent of water impairment under the Clean Water Act. For example, the US Environmental Protection Agency (EPA) recognizes iron as a source of water impairment. Water impairment data for States and Tribal Nations are available at https://mywaterway. epa.gov. For comparison, consider surface water contamination by iron in Illinois and West Virginia. While Illinois is approximately 2.4 times larger (by area), West Virginia has almost 12 times more river miles impaired by iron. As a result, it's necessary to understand the background levels of water quality constituents in order to place information into context.

In the US, the EPA maintains a list of impaired waters that are known as Clean Water Act 303(d) listed waters. Causes of impairment include a wide range of factors such as pathogens, nutrients, mercury, non-mercury metals, sediment, oil and grease, and polychlorinated biphenyls (PCBs), among others (US EPA 2022a). These data are collected as part of a larger effort to promulgate total maximum daily load (TMDL) criteria to aid states in better protecting their waters by developing load-based discharge regulations in contrast to concentration-based criteria. While this effort has not been uniform across every state, the data can provide valuable insights into major regional water impairment issues (US EPA 2022b).

REFERENCES

Chase, M. (1998). "NIST-JANAF Thermochemical Tables, Fourth Edition," *Journal of Physical and Chemical Reference Data, Monograph*, 9, 1951 pp.

Clark, A. (2021). "Karst Aquifers," US Geological Survey, www.usgs.gov/mission-areas/ water-resources/science/karst-aquifers, accessed July 20, 2021.

Doctor, D., and E. Alexander (2021). "Karst Geology of the Upper Midwest, USA," in *Caves and Karst of the Upper Midwest*, Springer Nature, Switzerland, 1–21 pp.

Doctor, D., Weary, D., Brezinski, D., Orndorff, R., and L. Spangler (2015). "Karst of the Mid-Atlantic Region in Maryland, West Virginia, and Virginia," *Field Guides*, 40, 425–484.

Egg Safety Center (2021). "What is the Safe Temperature to Cook Eggs?" https://eggsafety. org/faq/what-is-the-best-temperature-to-cook-an-egg/, accessed June 20, 2021.

NASA (2020). *Deserts*, National Aerospace and Aeronautical Administration, www.earth-data.nasa.gov/topics/biosphere/ecosystems/terrestrial-ecosystems/deserts, accessed November 24, 2020.

NIST (2021). *NIST Chemistry WebBook*, National Institute for Standards and Technology, https://webbook.nist.gov, accessed January 5, 2021.

US EPA (2022a). *Impaired Water and TMDLs, Overview of Listing Impaired Waters under CWA Section 303(d)*, US Environmental Protection Agency, www.epa.gov/tmdl/over view-listing-impaired-waters-under-cwa-section-303d#, accessed August 31, 2022.

US EPA (2022b). *Waters and Total Maximum Daily Loads (TMDLs)*, US Environmental Protection Agency, www.epa.gov/tmdl, accessed March 30, 2023.

Vasavada, A., Bandfield, J., Greenhagen, B., Hayne, P., Siegler, M., Williams, J., and D. Paige (2012). "Lunar Equatorial Surface Temperatures and Regolith Properties from the Diviner Lunar Radiometer Experiment," *Journal of Geophysical Research*, 117.

World Meteorological Organization (2016). "World/Global Weather and Climate Extremes Archive," http://wmo.asu.edu, accessed August 16, 2016.

3 Distribution and Cycling of Water on Earth

3.1 THE DISTRIBUTION OF WATER ON EARTH

The Earth is a water planet. Approximately 71% of the Earth's surface is covered by water, as presented in Figure 3.1 (Shiklomanov 1993). About 97% of this water is contained in the oceans, with the balance distributed in rivers, lakes, glaciers, and the ground. Since ~99% of freshwater (water with less than 1 g/L dissolved solids) is contained in glaciers and groundwater, only 1% is immediately accessible to humans in rivers, streams, lakes, and wetlands. When shown as percentages, it's easy to underestimate the scarcity of accessible freshwater. For example, for every liter of water on Earth, 970 mL is saline while only 30 mL is freshwater. Of the 30 mL of freshwater, only 0.3 mL is accessible as surface water.

Water withdrawals in the United States from 1955 to 2015 are presented in Table 3.1. According to the US Geological Survey (USGS), a total of 322,000 million gallons of fresh and saline water were extracted each day in the US in 2015 (Dieter *et al.* 2015). Of this total, 281,000 million gallons per day were freshwater withdrawals. This small fraction of the world's water is home to a little over 50% of all fish species, which amounts to nearly one quarter of the world's vertebrate species.

3.2 THE HYDROLOGIC CYCLE

Water is constantly moving between environmental compartments (*e.g.*, the atmosphere, oceans, soils, *etc.*) and/or changing form (as a gas, liquid, or solid). The movement of water through environmental compartments is represented schematically by the hydrologic cycle, which is presented in Figure 3.2. Major mechanisms of water movement through the environment include: precipitation, runoff (overland flow), percolation, infiltration, interflow, and groundwater flow.

3.2.1 PRECIPITATION

Precipitation includes all forms of water that are released from the atmosphere. This includes rain, snow, sleet, and hail. Water in the atmosphere is constantly evaporating and condensing. Liquid water is continuously being transformed into water vapor through evaporation. As water vapor is exposed to temperatures in the atmosphere that are lower than land temperatures, it condenses into small drops of liquid water. These droplets are too small to fall under the force of gravity. However, they are big enough to be seen as clouds, which are supported by the updraft of wind. For precipitation to form, condensed water droplets need to coalesce around a nucleus (small dust particles, smoke, or salts in the atmosphere). As additional water droplets collide

DOI: 10.1201/9781003289630-4

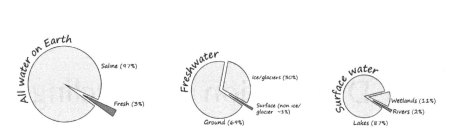

FIGURE 3.1 The distribution of water on Earth.

TABLE 3.1

Water Withdrawals in the United States from 1955 to 2015 (Dieter *et al.* 2015).

	1955	1975	1995	2015
Population, million people	164.0	216.4	276.1	325.0
Total water use, billion gal/d	240	420	398	322
Public water supply, billion gal/d	17.0	29.0	40.2	39.0
Rural domestic and livestock, billion gal/d	3.6	4.9	5.7	5.3
Irrigation, billion gal/d	110	140	130	118
Thermoelectric power cooling, billion gal/d	72	200	190	133
Other (aquaculture, mining, and self-supplied industrial), billion gal/d	39	45	31.35	26.35

Note: The sum of individual water uses may not equal the total due to rounding errors.

FIGURE 3.2 The hydrologic cycle is a schematic representation of the ways water travels between, within, and through environmental compartments.

with the nucleus, the mass of water becomes high enough to overcome the updraft velocity and rainfall begins.

3.2.2 PERCOLATION AND INFILTRATION

The processes of percolation and infiltration are often used interchangeably. However, they are distinctly different. Infiltration occurs when surface water enters the soil column at the soil surface, while percolation refers to the movement of water in the soil column. The rate of percolation depends on the soil water content and the hydraulic conductivity, K_s. Hydraulic conductivity is a measure of the relative ease with which water moves through a saturated porous media. It depends on factors that include the size, shape, and distribution of soil particles, as well as soil texture, roughness, and tortuosity (a measure of the twists and turns that are taken by water as it flows through soil pores). In a soil that is primarily composed of small clay particles, the void spaces between particles are correspondingly low and K_s values of around 10^{-10} m/s are typical. In contrast, soils that have relatively large void spaces such as mixtures of sand and gravel will have K_s values on the order of 10^{-4} to 10^{-2} m/s.

3.2.3 RUNOFF AND INTERFLOW

When precipitation hits an impervious surface or a saturated soil, gravity will carry it overland to a lower elevation. The extent of runoff depends on a number of physical, meteorological, and human factors that are summarized in Table 3.2. As water moves down-grade, soil, vegetation, and other materials are eroded and move with the bulk flow of water in a process known as advection. In this regard, many of the factors that influence runoff also have a direct impact on erosion. The extent to which these physical, meteorological, and human factors can impact runoff is also interrelated. For example, the impact of the time and intensity of the last precipitation event on soil saturation will also depend on the soil type and properties. Interflow can be thought of as subsurface runoff, as it's the lateral movement of subsurface water. Interflow occurs in the unsaturated zone (vadose zone)—between the soil surface and the water table. When compared to percolation and infiltration, interflow is more rapid.

3.2.4 EVAPOTRANSPIRATION

Evapotranspiration is the combined processes of evaporation and transpiration that results in the movement of water from land to the atmosphere from soil and vegetation. Evaporation occurs when liquid water changes to water vapor on soil and/or plant surfaces. Transpiration refers to the movement of water through plant leaves. Major factors that affect evapotranspiration include air temperature, humidity, wind speed, soil moisture, and the abundance of photosynthetically active radiation. Vegetation-specific factors can also impact the rate of evapotranspiration. For example, cacti and succulents transpire at a slower rate to conserve water. On a landscape that's covered with vegetation throughout the growing season, transpiration is

TABLE 3.2

Factors that Affect Runoff.

Physical Factors	Meteorological Factors	Human Factors
Soil type (sand, silt, clay) and properties (porosity, hydraulic conductivity)	Type of precipitation (rain, hail, snow, or ice)	Changes in land use (urbanization, draining agricultural fields, *etc.*)
Land use	Precipitation intensity (amount and duration)	Construction of manmade impervious surfaces
Land cover	Dew point	Deforestation by logging and other extractive industries
Slope/topography	Wind speed and direction	Human development activities, in general
Drainage area	Temperature	
The presence of lakes or reservoirs*	Time since and intensity of the last precipitation event	Construction of manmade lakes or reservoirs

* These structures can stop or slow the downstream movement of runoff.

the dominant water transport process. In this case, the movement of water is directly related to plant growth. When plants receive a sufficient supply of water and high air temperatures, stomates (the pores on plant leaves) open fully and both transpiration and photosynthesis occur at their highest rates. When water becomes limiting or the air temperature decreases, stomates close and the rates of both transpiration and photosynthesis decrease.

3.2.5   GROUNDWATER FLOW

Since around 69% of freshwater exists as groundwater (see Figure 3.1), the link between surface and subsurface water cannot be neglected. When surface water percolates into the soil, it migrates down to the water table. Since water has both cohesive and adhesive properties, soil pores located immediately above the saturated zone can become filled with water to form a capillary fringe. In this case, the adhesive force that acts between water and the soil pores is greater than the cohesive force between water particles. Water is held in the soil pores by surface tension. The height of the capillary fringe is determined by a balance between the adhesive force and the force of gravity.

The rate at which groundwater flows is determined by the ease with which water moves through the subsurface rock (permeability) and the amount of open space that's available for water to flow (porosity). When groundwater encounters a less permeable layer of subsurface material (rock or low permeability soil such as clay), it pools and can move horizontally. Likewise, water can move vertically though fractures in subsurface rock. Depending on the characteristics of the soil and subsurface rock, groundwater can remain in the subsurface from hours to hundreds and thousands of years.

On initial inspection, it's easy to overlook the utility of this simple schematic when used as a qualitative tool to infer where waters acquire particular characteristics. For example, as water moves through environmental compartments, a variety of elements are dissolved through interactions with the atmosphere, soil, and rock. This is the origin of the major and minor ions found in waters across the globe that were presented in the previous chapter. This also provides a schematic roadmap that we can use as we consider the ways water moves between environmental compartments.

In general, groundwaters have higher concentrations of dissolved [mineral] constituents than surface waters. Similarly, deep groundwaters generally have higher concentrations of dissolved constituents than shallower or geologically younger waters. This reflects the important role hydraulic residence time plays on the ultimate characteristics and quality of water. For example, Quinn (1992) provided an updated estimate of the hydraulic residence times of the Great Lakes. In this case, the hydraulic residence times for the Great Lakes were found to range from several months to hundreds of years, whereas hydraulic residence times for Lake Erie and Lake Superior were estimated at ~2.7 and 173 years, respectively. Similarly, the accumulation of groundwater—particularly that is held in deep aquifers—might best be measured on a geological time scale. In many cases, groundwater reservoirs act as water "sinks" until being tapped for use in irrigation or as a source of domestic drinking water. Likewise, a variety of biological processes can affect changes in the quality/characteristics of natural waters. These processes also include the biological activity in the subsurface environment.

While residence times of hundreds of years may seem long, in some cases, human ingenuity combined with a high demand for water can have significant impacts on the environment. For example, the High Plains Aquifer (also known as the Ogallala Aquifer) lies beneath eight states and has a footprint of ~450,000 km^2 (174,000 mi^2). Water from this aquifer is primarily used to irrigate crops. In 2009, water extraction from the High Plains Aquifer was used to support a food and fiber industry valued at ~$20 billion USD per year. Prior to ~1950, water in this aquifer was thought to be limitless. However, the rate at which water is being drawn down exceeds the rate at which it's being replenished (Water Resources 2021).

3.3   THE WATER CYCLE IN THE NEWS

While the Earth is a water planet, access to freshwater continues to be a challenge for many people. For example, global climate change has continued to cause rapid declines in the abundance of water across the globe. In 2022, waterborne commerce on the Rhine River in Germany was adversely affected by record-low water levels (Wittels *et al.* 2022). Likewise, the low water level on the Yangtze River caused the Chinese government to limit irrigation and power generation (Davidson 2022). In the United States, these same trends are reflected in a record 20+-year-long drought that's led to an unprecedented water shortage in the Lower Colorado River Basin (US Drought Monitor 2022; USBR 2012; Briscoe 2022). This water shortage has extended eastward to include the Mississippi River which experienced record-low water levels in October 2022 (US Drought Monitor 2022). This, too, has had an adverse impact on waterborne commerce at a time when farmers typically send their harvest to large ports to be shipped abroad (Nawaz 2022).

Throughout the first two decades of the 21st century, the western US experienced a significant drought that has had long-term hydrologic and ecological effects. Based

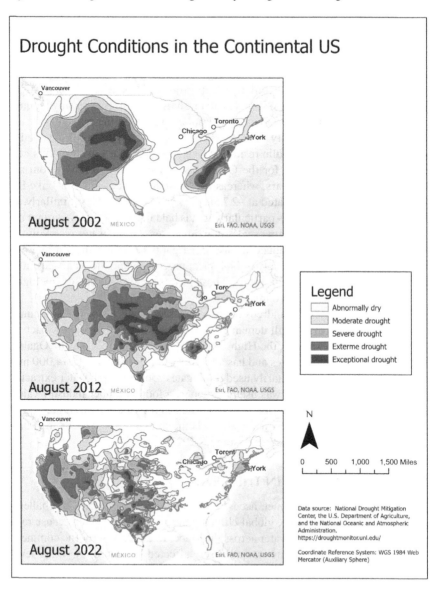

FIGURE 3.3   Drought conditions across the continental United States in 2002, 2012, and 2022.

on data from the National Drought Mitigation Center (NDMC), the states that make up the Colorado River Basin have been under drought conditions since at least 2000. Over the same period, other parts of the nation, including the Mississippi River Basin, also experienced drought conditions. To place this into context, the extent of drought conditions across the continental US in 2002, 2012, and 2022 are presented in Figure 3.3. In 2012, the severity of the drought was greater in the Upper and Lower Mississippi River Basins than in the Upper and Lower Colorado River Basins (Figure 3.3, middle). Fortunately, the severe drought along the Mississippi River Basin was relatively short-lived. However, drought conditions persisted in Minnesota, Wisconsin, Iowa, Illinois, Missouri, Arkansas, Kentucky, Tennessee, and Mississippi. While the duration and severity of the drought in the Mississippi River Basin are substantially less than those in the Colorado River Basin, the presence of drought conditions in the east had significant ecological and economic impacts. For instance, in late September 2022, a drought along the Mississippi River Basin caused low water conditions that had adverse impacts on waterborne commerce. Given the length and geographic extent of these dry conditions, water scarcity is likely to be a continual challenge in these chronically dry areas.

3.4 REFINING THE SCHEMATIC HYDROLOGIC CYCLE

The schematic hydrologic cycle presented previously in Figure 3.2 is a very general case. As we work toward developing an ability to describe the movement of mass in and between environmental compartments, we need to refine the graphic representation of the hydrologic cycle to better represent the bio-, hydro-, and geo-morphology of these systems. Let's begin by considering the schematic cross section of an idealized river corridor (Figure 3.4). In this system, the water level and degree of hydraulic connectivity are governed by a number of geomorphologic features including islands, levees, and river bluffs (FISRWG 1998).

Key factors that govern habitat development in these systems include the frequency of flooding, the elevation of a particular site relative to flood stage (low water level and flood stage in Figure 3.4), the proximity of the water table, and the soil porosity, among others. Initially, the establishment of floodplain forests is determined by the scouring and deposition of fertile, fine-grained deposits known as alluvial fill. The thickness of these deposits can vary from a few feet to over a hundred feet. After the establishment of these alluvial substrates and subsequent introduction/emergence of vegetation, habitats in bottomland forests are typically defined by a combination of elevation and the frequency of flooding.

Since the transition between habitats in bottomland forests is gradual, knowledge of geomorphic features can be useful in determining general habitat characteristics. As a result, additional work must be performed to delineate finer scale relationships between geomorphology and habitat. Consider low water conditions. In our schematic, the backwater lake (Figure 3.4) will generally have standing water. Intuitively, it should make sense that the backwater lakes provide a habitat that is suitable for submersed aquatic vegetation such as coontail (*Ceratophyllum demersum*) and water celery (*Vallisneria americana*; see Figure 3.5) (USDA 2020). In contrast, floodplain

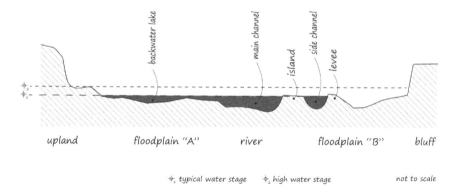

FIGURE 3.4 Schematic cross section of an idealized river corridor and key geomorphic features (FISRWG 1998).

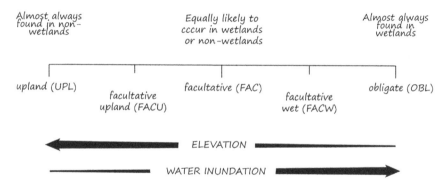

FIGURE 3.5 Classification of plants based on their likelihood to be found in wetland habitats (USDA 2020).

lakes flood intermittently and are inhabited by vegetation that is water tolerant but is also able to grow in the absence of standing water.

Within a single geomorphic feature, habitat changes along an elevation-based gradient. For example, in the floodplain lake in Figure 3.6, the habitat changes from shallow marsh to wet meadow to mesic prairie as elevation increases and the likelihood of encountering standing water decreases.

In large river systems, habitat is controlled by elevation. For example, shallow open water is typically found adjacent to the main channel. In Floodplain A in Figure 3.6, the backwater lake receives water from the main channel under typical hydrologic conditions. Under dry conditions, the backwater lake might not be directly connected to the main channel. The water elevation gradient within the backwater lake also influences the type of habitat that predominates. Hydrophytic vegetation is specially adapted to grow in very low oxygen conditions and is found in areas that experience prolonged water saturation. One way to classify plants is by the probability that they can be found in wetland habitats. This classification

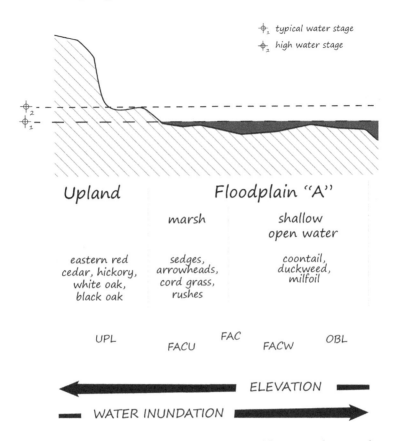

FIGURE 3.6　Relationship between vegetation, extent and frequency degree and extent of water inundation, and habitat type (after USACE 2020; USDA 2016).

scale ranges from obligate species including duckweed and coontail—those that are almost always found in wetlands and rarely in drier habitats—to upland species (Figure 3.6) (after FISRWG 1998; USACE 2020; USDA 2016).

As the elevation in Floodplain A increases and the soil is less fully and frequently inundated with water, marsh habitat develops (Figure 3.6). Cattail (*Typha*) and bulrush (*Scirpus*) are obligate plant varieties that are found in wet marshes. As elevation increases further, facultative wetland (FACW) species—those that generally occur in wetlands but can occasionally be found in non-wetlands—including prairie cordgrass (*Spartina pectinate*) are found. The US Army Corps of Engineers maintains the National Wetland Plant List (NWPL).[1] The US Department of Agriculture also provides a searchable database of wetland plants.[2]

3.5　THE WORLD'S LARGEST WATERSHEDS

The Amazon River, the Congo River, and the Mississippi-Missouri River systems are the world's three largest watersheds by area. Basin and biodiversity characteristics of the Amazon, Congo, and Mississippi-Missouri River watersheds are presented

TABLE 3.3

Basin Characteristics of the World's Three Largest [By Area] Watersheds (IUCN *et al.* 2003).

Basin Characteristic	Amazon R.	Congo R.	Mississippi R.
Area (km²)	6,145,186	3,730,881	3,202,185
Average population density (people/km²)	4	15	21
Number of cities with >100,000 people	16	18	20
Water supply per person (m³/person/ year) (1995 data)	273,767	22,752	8,973
Degree of river fragmentation	Medium	Medium	High
Number of dams in basin/on main stem	9	11	2,245
Number of dams >15 m (49.2 ft) high	8	11	2,091
Number of dams >150 m (492 ft) high	1	0	154
Number of fish species	3,000	700	375
Number of endemic fish species	1,800	500	85–127
Number of amphibian species	–	227	33
Number of Ramsar sites*	7	4	7
Number of endemic bird areas	24	6	0
Percent protected area	7.0	4.7	1.5
Protected area (km²)	430,163	175,351	48,033

* Wetlands of international importance as defined in the Ramsar Convention on Wetlands (1971) and subsequent amendments.

in Table 3.3 (IUCN *et al.* 2003). While the Mississippi-Missouri River system is the smallest of the three watersheds based on area, its population density is 40% greater than that of the Congo River and 425% greater than that of the Amazon River drainages.

In addition to its high population density, the *Mississippi-Missouri River* watershed is also characterized by a high degree of habitat fragmentation, which is reflected quantitatively in the number of dams located in the basin. While only ~7% of dams in the *Mississippi-Missouri River* drainage are classified as "high" (>150 m; 492 ft), the total number of dams eclipses the numbers found in the *Amazon* and *Congo River* drainages. It's also worthwhile to note that dams less than 15 m (49.2 ft) high aren't cataloged in Table 3.3. On the main stem of the *Mississippi River*, only one of 27 lower locks/dams (*Upper St. Anthony Falls*) has a vertical lift that exceeds 15 m (USACE 2016). Consequently, 26 discrete barriers to connectivity on the *Mississippi River* aren't accounted for in Table 3.3. The disparities among the three watersheds are even more apparent when comparing biodiversity characteristics. In particular, the inverse relationship between population density (Table 3.3) and the number of fish, bird, and amphibian species is clear. Additionally, when examining the total protected area in each basin, the impact of population density is further reinforced. For example, in the *Amazon River* basin, an area of 430,163 km² (166,087 mi²) is protected. In comparison, the total protected area in the *Mississippi-Missouri River* watershed is 48,033 km^2 (18,546 mi^2).

NOTES

1. See https://wetland-plants.usace.army.mil/nwpl_static/v34/home/home.html
2. See https://plants.usda.gov/home/wetlandSearch

REFERENCES

Briscoe, T. (2022). "As Talks on Colorado River Water Falter, U.S. Government Imposes New Restrictions," *Los Angeles Times*, www.latimes.com/environment/story/2022-08-16/colorado-river-basin-states-fail-to-reach-drought-agreement.

Davidson, H. (2022). "China Drought Causes Yangtze to Dry Up, Sparking Shortage of Hydropower,"*TheGuardian*,August22,2022,www.theguardian.com/world/2022/aug/22/china-drought-causes-yangtze-river-to-dry-up-sparking-shortage-of-hydropower.

Dieter, C., Maupin, M., Caldwell, R., Harris, M., Ivahnenko, T., Lovelace, J., Barber, N., and K. Linsey (2018). "Estimated Use of Water in the United States in 2015," *US Geological Survey Circular*, 1441, 65, https://doi.org/10.3133/cir1441.

Evers, Robert (1955). "Hill Prairies of Illinois," *Bulletin of the Illinois Natural History Survey*, 26 (5), 96.

FISRWG (1998). "Stream Corridor Restoration: Principles, Processes, and Practices," Federal Interagency Stream Restoration Working Group (FISRWG), GPO Item No. 0120-A; SuDocs No. A 57.6/2:EN3/PT.653., 637 pp.

International Union for Conservation of Nature (IUCN), International Water Management Institute (IWMI), the Ramsar Convention Bureau and the World Resources Institute (WRI) (2003). *Watersheds of The World, Water Resources eAtlas*, World Resource Institute, Washington, DC.

Little, Jane (2009). "The Ogallala Aquifer: Saving a Vital U.S. Water Source," *Scientific American*, May 1, 2009, www.scientificamerican.com/article/the-ogallala-aquifer.

McGuire, V. (2014). "Digital Map of Water-Level Changes in the High Plains Aquifer in Parts of Colorado, Kansas, Nebraska, New Mexico, Oklahoma, South Dakota, Texas, and Wyoming, Predevelopment (about 1950) to 2013," US Geologic Service, http://water.usgs.gov/lookup/getspatial?sir2014-5218_hp_wlcpd13.

Nawaz, Amna (2022). "Drought's Impact on Mississippi River Causes Disruptions in Shipping and Agriculture," *PBS News Hour*, November 17, 2022, www.pbs.org/newshour/show/droughts-impact-on-mississippi-river-causes-disruptions-in-shipping-and-agriculture.

Quinn, F. (1992). "Hydraulic Residence Times for the Laurentian Great Lakes," *Journal of Great Lakes Research*, 18 (1), 22–28.

Robertson, Ken. Personal observations made during field work between 1987 and 1996, www.inhs.uiuc.edu/~kenr/Revis.html, accessed September 1, 2014.

Schwegman, John. Unpublished species list based on field work in 1978 and 1979, Illinois Department of Natural Resources.

Shiklomanov, I. (1993). "World Fresh Water Resources," in Gleick, P. (Ed.), *Water in Crisis: A Guide to the World's Fresh Water Resources*, Oxford University Press, New York.

Taylor, J., Cardamone, M., and W. Mitsch (1990). *Bottomland Hardwood Forests: Their Functions and Values in Ecological Processes and Cumulative Impacts*, Illustrated by Bottomland Hardwood Wetland Ecosystems, Gosselink, J., Lee, L., and T. Muir (Eds.), Lewis Publishers, Inc., Chelsea, MI, 13–88 pp.

US Army Corps of Engineers (2016). www.mvr.usace.army.mil/Media/Fact-Sheets/, accessed August 16, 2016.

US Army Corps of Engineers (USACE) (2020). "National Wetland Plant List V.3.5, US Army Corps of Engineers," http://wetland-plants.usace.army.mil/.

US Bureau of Reclamation (2012). "Colorado River Basin Water Supply and Demand Study, US Bureau of Reclamation, US Department of the Interior," www.usbr.gov/watersmart/bsp/docs/finalreport/ColoradoRiver/CRBS_Executive_Summary_FINAL.pdf.

US Department of Agriculture, National Resource Conservation Service (USDA, NRCS) (2016). "The PLANTS Database, US Department of Agriculture, National Resource Conservation Service," http://plants.usda.gov.

US Drought Monitor (2022). "National Drought Mitigation Center, US Department of Agriculture, and the National Oceanic and Atmospheric Administration," https://drought monitor.unl.edu.

US Environmental Protection Agency (USEPA) (2023). "Stream Corridor Structure," *Watershed Academy Web*, https://cfpub.epa.gov/watertrain/moduleFrame.cfm?parent_object_id=610.

US Geological Survey (USGS) (2021). "Water Resources, High Plains Aquifer," *US Geological Survey*, www.usgs.gov/mission-areas/water-resources/science/high-plains-aquifer#overview.

Wittels, J., Kwaku, G., and L. Malsch (2022). "Europe's Rhine River is on the Brink of Effectively Closing," *Bloomberg.com*, www.bloomberg.com/news/articles/2022-08-02/europe-s-vital-rhine-river-is-on-brink-of-effectively-closing?leadSource=uverify%20wall.

4 Characteristics of Water

4.1 CHEMICAL CHARACTERISTICS

4.1.1 ELECTRICAL CONDUCTIVITY, IONIC STRENGTH, AND IONIC ACTIVITY

Electrical conductivity (EC) is a measure of a solution's ability to move electrons. EC is a function of the concentration of ions in solution, the valence charge of those ions, and factors such as temperature that impact the mobility of ions in solution. The specific conductance (SC) is a more standardized measure of electrical conductivity—the electrical conductance of 1 cm³ of a solution at 25°C (Hem 1982). Specific conductance is measured using a conductivity meter and has units of µS/cm. The specific conductance of a variety of waters is presented in Table 4.1 (USGS 2019).

Ionic strength, μ, is another gross measure of the abundance of ions in a solution. The concept of ionic strength was introduced in 1921 by Lewis and Randall, who offered a relationship to calculate μ in terms of the molar concentrations and respective charges of ions in a solution (Equation 4.1).

$$\text{Equation 4.1} \quad \mu = \frac{1}{2}\Sigma C_i Z_i^2$$

where C_i = molar concentration of ion i and z_i = the valence charge of the ith ion. Note the strong dependence of ionic strength on the charge of each ion.

One empirical relationship between ionic strength and specific conductance was given by Russell (1976; Equation 4.2). This relationship was developed based on samples from waters with broadly different characteristics and, as a result, can be applied to a range of natural waters.

$$\text{Equation 4.2} \quad \mu = \left(1.6 \times 10^{-5}\right) SC$$

where SC is the specific conductance in µS/cm.

An understanding of the relationship between ionic strength and the concentration and charge of major ions in a solution can be used to determine if all major ionic species have been accounted for in the analysis of a water sample. Ionic strength and specific conductance can be measured using a conductivity meter. If the measured and calculated values are different, there might be one or more ions that have not been included in the analytic work.

In dilute solutions, ions behave independently of one another. However, in higher ionic strength solutions, ions interact. When the number of ions in solution increases, the distance between ions decreases. This results in a portion of the system's total energy being expended by electrostatic interactions between ions, leading to an active concentration that's lower than the molar concentration by a factor γ (Equation 4.3).

$$\text{Equation 4.3} \quad \{C\} = \gamma [C]$$

DOI: 10.1201/9781003289630-5

TABLE 4.1

Specific Conductance of a Variety of Waters (USGS 2019).

Water	Specific Conductance, µS/cm
Rain	2–100
Freshwater streams	< 2,000
Ocean water	~50,000
Acid mine drainage	300–500,000
Wetlands	50–50,000

where {C} = active concentration of C, γ = activity coefficient, and [C] = molar concentration of C. As a general guideline, interactions between ions can become significant at solute concentrations in excess of ~10^{-4} M.

The activity coefficient can be determined using the Güntelberg approximation of the DeBye-Hückel theory (Equation 4.4) in which γ is related to ionic strength µ and the valence charge of an individual ion (Stumm and Morgan 1996; Debye and Hückel 1923). Since the ion charge is squared, we expect that the impact of ionic activity will be much greater in ions with a higher valence charge.

$$\text{Equation 4.4} \quad -\log(\gamma) = \frac{1}{2}Z^2 \frac{\sqrt{\mu}}{1+\sqrt{\mu}}$$

Example: The Role of Ionic Strength in Metal Toxicity toward Aquatic Receptors

Many waterbodies are impacted by industrial discharges of clean water—water that's been treated, but still contains high concentrations of certain constituents. One example is acid mine drainage treatment, where water is typically aerated to oxidize Fe^{2+} to Fe^{3+} then dosed with a chemical (often $Ca(OH)_2$) to increase the pH. This creates the chemical conditions needed to precipitate ferric iron and many other metals from solution as metal hydroxides/oxides.

It's not uncommon to see trout living and growing in streams that are immediately down-grade from the discharge of these treatment facilities. Given the sensitivity of salmonids to dissolved metals, this observation can be confounding.

Consider a channel located downstream from the regulated discharge from an acid mine drainage treatment facility. Representative characteristics of the water are presented in Table 4.2. The pH is circumneutral and the water temperature is generally hospitable for trout. However, dissolved metal concentrations reflected by the high specific conductance can have adverse impacts on freshwater fish. In this case, recommended maximum concentrations of dissolved iron, aluminum, and magnesium for salmonids are 0.1, 0.01, and 15 mg/L, respectively (Meade 1998). In this case, the metal concentrations presented in Table 4.2 exceed maximum recommendations by as much as a factor of 10. Under these conditions, some signs of fish stress would be expected but were not observed. Why? Equation 4.5 is just the application of equation 4.2.

TABLE 4.2

Partial Analysis of Water Downstream from an Acid Mine Drainage Treatment Facility.

Parameter	Value
pH	7.2
Temperature	13 °C
Specific conductance	6.7×10^3 μS/cm
Dissolved iron	0.20 mg/L
Dissolved aluminum	0.13 mg/L
Dissolved magnesium	121 mg/L

The data in Table 4.2 represent a partial analysis, so we can't be sure we've accounted for all ions in solution. However, since we know the measured specific conductance, we can estimate the ionic strength using Equation 4.5.

$$\text{Equation 4.5} \quad \mu = \left(1.6 \times 10^{-5}\right)\left(6.7 \times 10^3\right) = 0.1072 \cong 0.11$$

Based on this ionic strength, we can determine the activity coefficients for the ions in our stream water. If we assume the water is well aerated, iron will be present as Fe^{3+}. Aluminum is a trivalent ion and magnesium is a divalent ion. Consequently, we need to determine the effect of elevated ionic strength on di- and trivalent ions. After a little arithmetic (solving Equation 4.4 for γ), you'll find that the activity coefficients are 0.32 and 0.07 for di- and trivalent ions, respectively. In practical terms, this means that for divalent ions in the stream water, the active concentration will only be 32% of the measured concentration. Likewise, the active concentrations of trivalent ions are reduced by 93%. In the case of trivalent ions, the active concentrations of iron and aluminum are 0.014 mg/L and 0.009 mg/L, respectively. These concentrations are below recommended maximum concentrations. While the active concentration of magnesium (39 mg/L) still exceeds recommended limits, major stresses induced by the trivalent ions have been alleviated.

4.2   ALKALINITY AND ACIDITY

Alkalinity is a measure of water's ability to resist changes in pH when exposed to a strong acid (an acid with a high tendency to release protons). Similarly, acidity is the ability of water to resist a change in pH when a strong base is added. Strong acids are generally introduced into water as salts such as H_2SO_4 and HCl, and strong bases such as NaOH are often introduced as hydroxides/oxides.

We're going to develop a few different ways to determine alkalinity based on the results of different known factors. No one method for determining alkalinity is better than any other. Rather, one approach might simply be more convenient given what information is known.

Worldwide, natural waters typically contain some or all of the following dissolved ions: Na^+, K^+, Ca^{2+}, Mg^{2+}, Cl^-, SO_4^{2-}, H^+, OH^-, HCO_3^-, and CO_3^{2-}. We can write a general electroneutrality equation in terms of the equivalent concentrations of these constituents (Equation 4.6).

$$\text{Equation 4.6} \quad \left[Na^+\right]+\left[K^+\right]+2\left[Ca^{2+}\right]+2\left[Mg^{2+}\right]+\left[H^+\right]$$
$$=\left[OH^-\right]+\left[Cl^-\right]+\left[HCO_3^-\right]+2\left[SO_4^{2-}\right]+2\left[CO_3^{2-}\right]$$

Since we're balancing charge, care should be taken with terms representing multi-valent ions such as Ca^{2+}, Mg^{2+}, SO_4^{2-}, and CO_3^{2-} in Equation 4.6, as each of these species contain two equivalent units of charge per mole. Notice that we can view Equation 4.6 as a combination of acids, bases, and carbonate species. You might be wondering why we're separating the carbonate species since atmospheric carbon dioxide acts as a weak acid when it enters the water phase. In this case, we're considering soluble carbonate species (CO_2 (aq), H_2CO_3 (aq), HCO_3^-, CO_3^{2-} (aq)) separately for convenience, as you'll see in the next steps. If we define C_B as the charge of strong base cations and C_A as the charge of strong acid anions, we can express our charge balance by writing Equation 4.6 in a slightly different form (Equation 4.7).

$$\text{Equation 4.7} \quad C_B +\left[H^+\right]=C_A +\left[OH^-\right]+\left[HCO_3^-\right]+2\left[CO_3^{2-}\right]$$

Since alkalinity is a measure of a water's ability to resist a change in pH when a strong acid is added, in the context of Equation 4.7, the difference between C_B and C_A is the acid neutralizing capacity (ANC). In most natural waters, ANC is determined by carbonate species, hydroxide ions, and protons. This allows us to write two expressions for alkalinity, as shown in Equation 4.8 and Equation 4.9.

$$\text{Equation 4.8} \quad Alk =\left[OH^-\right]+\left[HCO_3^-\right]+2\left[CO_3^{2-}\right]-\left[H^+\right]$$

$$\text{Equation 4.9} \quad Alk =\left(\left[Na^+\right]+\left[K^+\right]+2\left[Ca^{2+}\right]+2\left[Mg^{2+}\right]\right)-\left(\left[Cl^-\right]+2\left[SO_4^{2-}\right]\right)$$

While both expressions for alkalinity are always correct, one can be more useful than the other under certain circumstances. For example, in Equation 4.8, anything that accepts a proton will contribute to alkalinity. In contrast, when making up reagents or determining chemical requirements in water treatment, Equation 4.9 may be of greater utility.

4.3 HARDNESS

Total hardness, TH, is calculated as the sum of the equivalent concentrations of all polyvalent cations (*i.e.*, ions with a charge of 2+ or greater) as shown in Equation 4.10, where [Me^{n+}] is the molar concentration of a cation, Me, with n equivalents of charge per mole (n > 1).

$$\text{Equation 4.10} \quad TH = \sum n[Me^{n+}], n > 1$$

A general scale of water hardness is presented in Figure 4.1. On a worldwide basis, most total hardness is due to dissolved calcium and magnesium. However, other

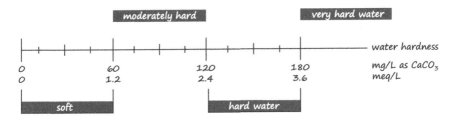

FIGURE 4.1   Relative scale of water hardness.

polyvalent cations including iron and aluminum can contribute to hardness. As a result, a complete analysis of water is needed to ensure that local conditions are considered.

Water hardness is also classified based on the characteristics of the anion with which Me^{n+} is associated. For example, carbonate hardness (CH) is derived from polyvalent cations associated with HCO^- and CO_3^{2-}. Examples include $Ca(HCO_3)_2$ and $CaCO_3$. In contrast, noncarbonate hardness (NCH) results from Me^{n+} associated with noncarbonate ions as well as free Me^{n+}.

The total hardness is the sum of carbonate and noncarbonate hardness, CH and NCH, respectively, as shown in Equation 4.11.

$$\text{Equation 4.11} \quad TH = CH + NCH$$

Example: Interpreting the Results of Hardness Testing

As a professional scientist, there's little chance that you're going to be solely responsible for collecting samples, analyzing them, and interpreting the data. Rather, you'll most likely rely on the assistance of technicians, chemists, and others to provide the data you need to inform your decisions. In that spirit, consider two water samples that were collected from a local creek and analyzed by a certified analytic laboratory. An analysis of major ions and alkalinity for both water samples is presented in Table 4.3. Determine the total hardness, carbonate hardness, and noncarbonate hardness of two water samples.

The first thing you might notice is that the water pH isn't specified. In this case, we need to assume the pH is circumneutral. For natural waters, this is a good starting point—especially if you're not given any other information. Recall that we have a number of choices for how we represent alkalinity and should use the one that best suits the context of our work. In this case, we need to extract some useful information related to water hardness from a known alkalinity. One approach is to use Equation 4.8, which relates alkalinity to the pH and carbonate species. At a neutral pH, $[H^+] \sim [OH^-]$, and $[HCO_3^-] \gg [CO_3^{2-}]$ so any carbonate alkalinity will be due to the presence of HCO_3^-.

We can find the total hardness of each sample using Equation 4.10 by adding the equivalent concentrations of the polyvalent cations, Ca^{2+} and Mg^{2+}. The total hardness of each sample is 6 eq/L. Chloride and sulfate ions are anions and don't contribute to hardness.

TABLE 4.3
An Analysis of Major Ions and Alkalinity of Two Creek Water Samples.

Constituent	Sample 1	Sample 2
Alk	7 eq/L	3 eq/L
Ca^{2+}	2 eq/L	5 eq/L
Mg^{2+}	4 eq/L	1 eq/L
Cl^-	4 eq/L	2 eq/L
SO_4^{2-}	1 eq/L	1 eq/L

Based on our earlier assumptions, any carbonate hardness will be due to the presence of bicarbonate. Water Sample 1 contains 7 eq/L of bicarbonate. In this sample, all 6 eq/L of total hardness will be associated with the bicarbonate ions. Consequently, the hardness in Sample 1 is all carbonate hardness and 1 eq/L of bicarbonate remains in solution. In contrast, Sample 2 contains 3 eq/L of bicarbonate. While Sample 2 also has 6 eq/L of total hardness, it has 3 eq/L of carbonate hardness and 3 eq/L of noncarbonate hardness.

In municipal water supplies, elevated hardness (> 2.4 meq/L as in Figure 4.1) leads to aesthetic issues such as difficulty in rinsing soap from surfaces and the formation of a film/scale on fixtures. In industrial applications, the scale that develops on the surface of heat exchangers can decrease the life of equipment and increase the frequency, extent, and cost of operation and maintenance.

In natural systems, elevated water hardness can be an indication of the presence of potentially toxic dissolved metals. Likewise, as we've seen, the presence of certain compounds in water, including carbonate species that are derived from $CaCO_3$, can increase alkalinity. This helps to increase a stream's ability to resist changes in pH. Subsequent additional impacts include changes to metal solubility that could favor the formation or dissolution of metal oxide solids. While the formation of precipitates results in the removal of dissolved metals from the water column, the solids formed in this process can adversely impact organisms that utilize stream beds as habitat.

4.4   OXYGEN DEMAND

When characterizing water quality, gross measures based on the concept of oxygen demand are frequently used as indicators of the presence of organic compounds. Metrics including biochemical and chemical oxygen demand reflect the total concentration of organics in water, but not the specific chemical forms of those compounds. Consider the schematic equation for the aerobic degradation of organic matter (OM) in Equation 4.12.

Equation 4.12 $OM + O_2 + nutrients \rightarrow H_2O + CO_2 + cells + energy + byproducts$

Based on Equation 4.12, we can define biochemical oxygen demand (BOD) as the amount of oxygen used in the aerobic metabolism of biodegradable organic matter.

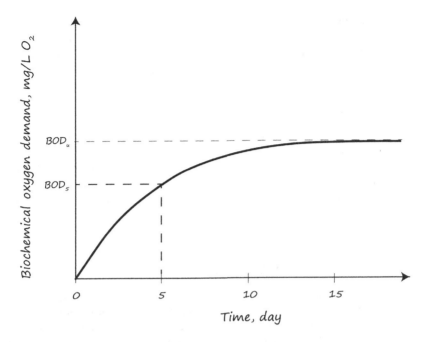

FIGURE 4.2   A schematic plot of biochemical oxygen demand (BOD) versus time.

A schematic plot of BOD over time is presented in Figure 4.2. Since the BOD test is based on a biochemical reaction that takes time to occur, a period of five days is used to standardize the analysis (BOD_5). Even though this five-day period is used as a standard time for the BOD test, aerobic microorganisms can continue to metabolize organic matter beyond five days. When all of the soluble oxidizable organic matter in a sample is completely metabolized by the aerobic microorganisms, the BOD concentration no longer changes over time. This is known as the ultimate BOD, or BOD_u (see Figure 4.2).

Depending on the composition of organic matter and the degree to which dissolved oxygen is utilized, the total BOD is made up of carbonaceous and nitrogenous components. Carbonaceous BOD is the amount of oxygen used to degrade organic compounds that either don't contain nitrogen or in which nitrogen is converted to NH_3 with no additional dissolved O_2 consumption. Any oxygen that's used to degrade NH_3 by nitrogen oxidizing bacteria (NOB) is known as nitrogenous BOD. Ammonia is aerobically metabolized to create nitrate in a two-step process. First, ammonia is oxidized to nitrite by *Nitrosomonas* (Equation 4.13), and then followed nitrite is oxidized to nitrate by *Nitrobacter* (Equation 4.14).

$$\text{Equation 4.13}\quad NH_3 + \frac{3}{2}O_2 \rightarrow NO_2^- + H^+ + H_2O$$

$$\text{Equation 4.14}\quad NO_2^- + \frac{1}{2}O_2 \rightarrow NO_3^-$$

In domestic wastewaters, it can take 5–8 days before NOB begin to oxidize nitrogen-containing compounds. The NOB bacteria in Equation 4.13 and Equation 4.14 are both chemo-litho-autotrophic (their only energy source is chemical energy; inorganic carbon is their electron donor and CO_2 is their carbon source). These equations can be added algebraically to obtain the overall reaction in Equation 4.15.

$$\text{Equation 4.15} \quad NH_3 + 2O_2 \rightarrow NO_3^- + H^+ + H_2O$$

The rate at which nitrification occurs is a function of several environmental factors, including the concentration of organic matter and dissolved oxygen, temperature, pH, and the presence of rate-limiting compounds (particularly those that are toxic to NOB). These reactions also provide an important pathway for nutrient utilization in aquatic and terrestrial ecosystems. Through this process, plants receive the nitrates needed to form amino acids and subsequently create the proteins needed for cell growth.

4.5   CHEMICAL OXYGEN DEMAND

Chemical oxygen demand (COD) is another measure of the amount of organic material in water. The COD test is based on the amount of oxygen consumed during oxidation. However, in COD analysis, a strong chemical oxidizer is used in place of oxidizing bacteria. In COD testing, a sample is mixed with a strong oxidizer and acid. The solution is heated, and the oxygen demand is determined based on the amount of reagent needed to reach a particular titrimetric or photometric endpoint. It's important to note that other reduced species (*e.g.*, sulfides, sulfites, and ferrous iron) will be oxidized and can contribute to COD.

In contrast to BOD testing, COD tests generally take less than three hours to complete. Additionally, COD testing is not adversely affected by the presence of compounds such as dissolved metals that could be toxic to NOB. Due to the relatively short time of analysis, BOD tests are often supplemented by interim measurements of COD. When supplementing BOD_5 data with COD testing, a correlation between the two must first be established. Likewise, the characteristics of biochemically oxidizable organics in water should be relatively constant.

If we know the BOD and COD, we can draw several important conclusions about a water. For example, when BOD and COD are equal, the organic matter in water is completely biodegradable. If COD > BOD, we can estimate the degree to which the water is biodegradable as the ratio of BOD to COD.

Example: Determining Biodegradability from Analytic Data

Consider water with a COD of 1,200 mg/L and a BOD_5 of 1,188 mg/L. Estimate the percentage of nonbiodegradable organic material in the water.

COD is the concentration of all oxidizable organic material and BOD_5 is an estimate of the concentration of biodegradable material, so the difference is the nonbiodegradable concentration of organic material. In this case, that's 12 mg/L. As a fraction of the total degradable organic material in the water (COD), the nonbiodegradable fraction is about 1% ((12/1,200)100).

Example: Relating Oxygen Demand to Source Streams

It's Friday afternoon at 4:59 pm and a member of your team just finished analyzing water samples for BOD_5 and COD. A summary of the data was slipped under your door (see Table 4.4). As a dedicated employee, you decide to review the data before leaving for the weekend. The samples include a wastewater that contains mostly biodegradable organics, water from a seemingly clear stream, a wastewater that is known to contain nonbiodegradable organics, and a sample that someone added to the analytic queue without any documentation. Based on the information in Table 4.4, identify each sample and explain your reasoning.

TABLE 4.4
Summary of BOD_5 and COD Data.

Sample No.	BOD_5, mg/L	COD, mg/L
Limit of detection (LOD)	2	1
001	1,175	1,200
002	4	5
003	1,750	1,250
004	125	2,500

- Sample 1 exerts a nonnegligible oxygen demand. The organic material in this sample is 97.9% biodegradable ((1,175/1,200)100 = 97.9%). This is probably the wastewater that contains biodegradable organics.
- The oxygen demand in Sample 2 is close to the limit of detection. This sample has the lowest oxygen demand of the four samples. Sample 2 is most likely from the seemingly clear stream.
- In Sample 3, BOD_5 is greater than the COD. Clearly, something is wrong with this sample. From a practical point of view, the concentration of biochemically oxidizable organic material (BOD) will never exceed the oxygen demand exerted when a strong chemical oxidizer is applied (COD).
- Sample 4 exerts the greatest total oxygen demand. However, only 5% of the soluble organic material in the sample is biodegradable ((125/2,500)100 = 5%). Sample 4 is most likely the industrial wastewater that contains a low concentration of soluble organic material.

4.6 PHYSICAL CHARACTERISTICS OF WATER

Temperature, clarity, and solids concentration are among the most important physical parameters that impact the way constituents interact in aquatic systems.

4.6.1 TEMPERATURE

According to the USGS, ~48% of water withdrawals in the US in 2000 were made to cool water in the production of thermoelectric power. This figure doesn't include water used in hydroelectric power generation. In terms of its magnitude,

TABLE 4.5

Solubility of Oxygen in Freshwater at 1 atm (101.1 kPa) Over a Range of Environmentally Relevant Water Temperatures (Haynes 2014).

Water Temperature, °C	Dissolved Oxygen Concentration, mg/L
0	14.6
5	12.8
10	11.3
15	10.1
20	9.1
25	8.3
30	7.6
35	7

135,000 Mgal/d of fresh water and 59,500 Mgal/d of saline water were extracted for this one application.

To understand the significance of this fraction of water withdrawal in the US, it's worthwhile to compare water withdrawal for thermoelectric power cooling with that used for irrigation in states with large agricultural production. For example, in Illinois, 11,300 Mgal/d of fresh water were withdrawn and used for the production of electricity in 2000. This accounted for ~74% of total water withdrawal in the state. In contrast, during the same time period, only 4.25 Mgal/d were extracted for use in irrigation (Hutson *et al.* 2004).

The significance of water temperature on aquatic receptors cannot be overemphasized. As the temperature of water increases, the solubility of oxygen in water decreases (see Table 4.5). At the same time, warmer water temperatures can cause aquatic organisms to increase respiration rates, further exacerbating the impact of the decreased oxygen solubility. These conditions increase overall stress on aquatic receptors, which can leave them susceptible to opportunistic pathogens. Elevated water temperature can also disrupt the growth and reproductive cycles of aquatic organisms. For example, Pankhurst and Munday (2011) reported on the impacts of climate-driven changes in temperature on fish reproduction. They found that higher temperatures resulted in shorter spring spawning cycles, while fall spawning was delayed by higher water temperatures. They attributed these observations to the sensitivity of larval fishes to changes in water temperature.

4.6.2 Clarity and Color

Turbidity is one indicator of water clarity. Specifically, turbidity is a measure of the amount of colloidal-sized particles (very small solids that remain dispersed in a medium for an extended period of time) in water. Secchi disks and derivative systems such as "turbidity tubes" are common tools used to measure turbidity. In each method, the visibility of a viewing disk is measured as water depth is increased. Using a depth-to-turbidity calibration, the turbidity of the water is determined based on the depth where the disk is no longer visible. These methods are typically used when taking measurements in ponds and lakes.

Turbidity can also be measured using an instrument in which a light source of known intensity is passed through a water sample contained in a cell of known geometry and material properties. A detector located opposite the light source measures the fraction of incident light that passes through the water sample. Turbidity is inversely proportional to the fraction of light transmitted and is measured in nephelometric turbidity units (NTU).

The color of water can come from suspended and/or dissolved constituents. In some cases, the color of a solution comes from the scattering of light by small particles. This is known as "apparent color". For example, the brownish color of Mississippi River is due to small alluvial material that remains suspended in the water column (see Figure 4.3, left). The turbidity of the Mississippi River sample in Figure 4.3 was 320 NTU. When the sample was filtered through a 0.21 um glass fiber filter, the small particles were removed (Figure 4.3, right), and the turbidity of the filtrate was < 1 NTU.

Color due to compounds that are dissolved in solution is referred to as "true color." In natural waters, true color typically comes from organic acids that occur when natural organic matter decomposes. The Blackwater River in central

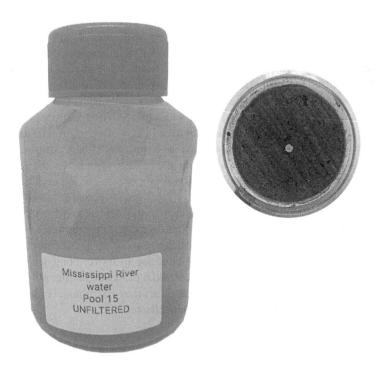

FIGURE 4.3 Mississippi River water collected from Pool 15 (Moline, IL, USA; left), fine alluvial material retained on a 0.21 um glass fiber filter (right). Photo by Roger C. Viadero, Jr., March 14, 2023.

West Virginia was named for the dark, over-steeped tea color of the water (see Figure 2.4). The color comes from humic and fulvic acids resulting from the decomposition of natural organic matter from the surrounding mixed forest. While these waters appear dark in color, turbidity is generally low. In contrast, color that is due to the presence of small suspended solids is known as "apparent color." Consequently, a sample with true color will have a [much] lower turbidity than a sample with apparent color.

4.7   SOLIDS

The total solids (TS) concentration in a water sample is the sum of the total dissolved solids (TDS) and total suspended solids (TSS) concentrations. In water analysis, the TS concentration is found gravimetrically by determining the mass of water evaporated from a sample of known mass after heating it in a lab oven at 103–105 °C. The TS concentration is the difference between the steady-state masses of the sample before and after heating divided by the known sample volume. The full procedure to determine the TS concentration is given by the American Public Health Association (APHA 2005a). Total dissolved and total suspended solids concentrations are found in a similar manner after passing the sample thorough a filter of known mass (APHA 2005b).

These procedures are conducted in the laboratory and can be time consuming. TDS concentration is often approximated using the electric conductivity since dissolved salts frequently represent a large fraction of the dissolved solids in many waters. Likewise, turbidity is often used as a field-based indicator of the TSS concentration. In addition to affecting the aesthetics of water (odor and color), elevated TS and TSS concentrations can be an indicator of the presence of small organic particles that are known to serve as host sites for opportunistic pathogens.

The concentration of volatile suspended solids (VSS) (APHA 2005c) is found by measuring the steady-state mass of the filter used in TSS analysis after heating it in a muffle furnace at 550 °C. The difference in mass before and after heating at 550 °C is proportional to the mass of fixed solids, while the balance is related to the concentration of volatile solids. The VSS concentration is proportional to the organic fraction of suspended solids in a sample.

Particle size is a key factor used to determine the degree to which solids will tend to remain in solution. For particles of equal density, larger particles will tend to settle out of solution faster than smaller particles. As long as small particles remain suspended in the water column, they will continue to move along with the bulk flow of water. From a treatment point of view, the removal of smaller particles from water generally involves more advanced approaches than the removal of large solids, which will be covered in detail in a subsequent chapter.

REFERENCES

American Public Health Association (2005a). "Method 2540B Total Solids 209C Dried at 103–105°C," in *Standard Methods for the Examination of Water and Wastewater*, 21st ed., American Public Health Association, New York, 1496 pp.

American Public Health Association (2005b). "Method 2540 D Total Suspended Solids Dried at 103–105°C," in *Standard Methods for the Examination of Water and Wastewater*, 21st ed., American Public Health Association, New York, 1496 pp.

American Public Health Association (2005c). "Method 2540E Fixed and Volatile Solids Ignited at 550°C," in *Standard Methods for the Examination of Water and Wastewater*, 21st ed., American Public Health Association, New York, 1496 pp.

Debye, P., and E. Hückel (1923). "The Theory of Electrolytes. I. Freezing Point Depression and Related Phenomena" ["Zur Theorieder Elektrolyte. I. Gefrierpunktserniedrigung und verwandte Erscheinungen"], *Physikalische Zeitschrift*, 24, 185–206.

Flegal, T., and E. Schroeder (1976). "Temperature Effects on BOD Stoichiometry and Oxygen Uptake Rate," *Journal of the Water Pollution Control Federation*, 49 (12), 2700.

Hem, J. (1982). "Conductance — A Collective Measure of Dissolved Ions," in Minear, R., and L. Keith (Eds.), *Water Analysis—v. 1 Inorganic Species, Part 1*, Academic Press, Inc., Orlando, FL, 137–161 pp.

Hutson, S., Barber, N., Kenny, J., Linsey, K., Lumia, D., and M. Maupin (2004). *Estimated Use of Water in the United States in 2000, US Geological Survey Circular 1268*, US Geological Survey, Reston, VA, 52 pp.

Ingraham, J. (1959). "Growth of Psychrophilic Bacteria," *Bacteriology*, 76 (1), 75.

Lewis, G., and M. Randall (1921). "The Activity Coefficients of Strong Electrolytes," *Journal of the American Chemical Society*, 43, 1112–1154.

Meade, J. (1989). *Aquaculture Management*, Van Nostrand Reinhold, New York, 190 pp.

National Institute for Standards and Technology (NIST) (2021). *NIST Chemistry WebBook*, US Department of Commerce, National Institute for Standards and Technology, https:// webbook.nist.gov, accessed January 5, 2021.

Pankhurst, N., and P. Munday (2011). "Effects of Climate Change on Fish Reproduction and Early Life History Stages," *Marine and Freshwater Research*, 62, 1015–1026.

Russell, L. (1976). "Chemical Aspects of Groundwater Recharge with Wastewaters," Ph.D. Thesis, University of California, Berkeley.

Sanders, L. (1998). *A Manual of Field Hydrogeology*, Prentice-Hall, Upper Saddle River, NJ, 381 pp.

Stumm, W., and J. Morgan (1996). *Aquatic Chemistry: Chemical Equilibria and Rates in Natural Waters*, 3rd ed., John Wiley and Sons, New York, 103–105 pp.

US Geological Survey (USGS) (2019). "Specific Conductance: US Geological Survey Techniques and Methods, Book 9," Chap. A6.3, 15, https://doi.org/10.3133/tm9A6.3.

Viadero, R., and A. Tierney (2003). "Use of Treated Mine Water for Rainbow Trout (Oncorhynchus mykiss) Culture—A Preliminary Assessment," *Aquacultural Engineering*, 29 (1–2), 43–56.

5 Chemical Reactions in Aquatic Systems

5.1 CHEMICAL EQUILIBRIUM

Consider the general single-step reversible chemical reaction presented in Equation 5.1 that describes the reaction of A and B to create C and D.

$$\text{Equation 5.1} \quad aA + bB \leftrightarrow cC + dD$$

where a, b, c, and d are stoichiometric coefficients, A and B are reactants, and C and D are products. Over time, A and B are consumed and C and D are created. Since we know that matter cannot be created or destroyed, elements that appear as reactants must also appear in equal molar amounts on the product side of the equation. Likewise, the net charge on each side of a balanced equation must be the same. At equilibrium, the reaction rate is constant over time.

It's important to realize that Equation 5.1 is actually a composite of a forward reaction (from reactants to products) and a reverse reaction (from products to reactants), which are respectively characterized by forward and reverse reaction rates with general forms given in Equation 5.2 and Equation 5.3.

$$\text{Equation 5.2} \quad r_f = -k_f \left[A\right]^a \left[B\right]^b$$

where r_f = forward reaction rate and k_f = forward reaction rate constant. If the unit of reaction rate is mol/time, the units of the equilibrium reaction constant will depend on the values of a and b in Equation 5.2.

$$\text{Equation 5.3} \quad r_r = -k_r \left[C\right]^c \left[D\right]^d$$

where r_r = reverse reaction rate and k_r = reverse reaction rate constant.

When a reacting chemical system reaches equilibrium, the forward and reverse reaction rates are constant. In this case, the reaction is characterized by the equilibrium reaction constant, K_{eq}. For a single-step reversible reaction at equilibrium, we can equate Equation 5.2 and Equation 5.3 and write the equilibrium constant, K_{eq}, in terms of product and reactant concentrations raised to their respective stoichiometric coefficients as in Equation 5.4.

$$\text{Equation 5.4} \quad K_{eq} = \frac{\left[C\right]^c \left[D\right]^d}{\left[A\right]^a \left[B\right]^b}$$

The equilibrium reaction constant is a function of environmental conditions, including temperature, pressure, and ionic strength.

DOI: 10.1201/9781003289630-6

5.2 CHEMICAL KINETICS

Previously, we considered the case where equilibrium was reached in a chemical reaction. Now, we need to examine the case where equilibrium between reactants and products hasn't been reached, so net changes in their respective concentrations occur over time. Recall the balanced chemical reaction presented previously in Equation 5.1. We can write a general expression for the rate at which the reaction occurs, r, according to Equation 5.5.

$$\text{Equation 5.5} \quad r = -k[A]^w[B]^x[C]^y[D]^z$$

where w, x, y, and z are coefficients determined through lab studies. The unit for r is mol/l·s while the units of k are a function of the values of w, x, y, and z.

The overall order of a reaction is determined by adding the coefficients in Equation 5.5. The order of a reaction with respect to an individual product or reactant is determined by the exponent associated with a particular species. In our general reaction rate expression, the reaction is "w order" in A, "x order" in B, "y order" in C, and "z order" in D.

5.2.1 ZERO-ORDER REACTIONS

Consider the experiment depicted in Figure 5.1 where "A" is added at t = 0 to a beaker at an initial concentration of $[A_0]$. No additional A is added. The beaker is completely mixed and [A] is measured over time. The general form of a zero-order reaction is presented in Equation 5.6. Recall that anything raised to the zero power is equal to one.

$$\text{Equation 5.6} \quad r = \frac{\Delta[A]}{\Delta t} = -k[A]^0 = -k$$

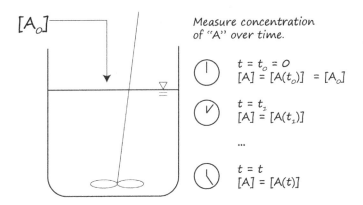

FIGURE 5.1 An experiment to determine the order of a reaction in a batch reactor.

In differential form, the rate equation for a zero-order reaction is given in Equation 5.7.

$$\text{Equation 5.7} \quad r = \frac{d[A]}{dt} = -k$$

We can find the change in the concentration of A over a time period, Δt, from Equation 5.8 for the discrete case and from Equation 5.9 in differential form.

$$\text{Equation 5.8} \quad \Delta[A] = -k\Delta t$$

$$\text{Equation 5.9} \quad \int_{[A_0]}^{[A]} d[A] = -k\int_0^t dt$$

By applying the following initial conditions, we can solve either Equation 5.8 or Equation 5.9 for the concentration of A as a function of time: at $t = 0$, $[A] = [A_0]$ and $[A(t)] = [A]$.

$$\text{Equation 5.10} \quad \left[A(t)\right] = [A_0] - kt$$

Equation 5.10 describes a line with a y-intercept of $[A_0]$ and a slope of k (Figure 5.2). The slope of the line is negative since we considered our reaction to be the consumption of "A" over time. What if the experimental data are not linear on a plot of concentration versus time? At this point, the only conclusion we can draw is that the data are not described by zero-order reaction kinetics.

5.2.2 FIRST-ORDER REACTIONS

Now, let's consider the general form of a first-order reaction (Equation 5.11).

$$\text{Equation 5.11} \quad r = \frac{\Delta[A]}{\Delta t} = -k[A]^1$$

In this case, we need to write the reaction rate in differential form (Equation 5.12). To start this solution, we need to gather like terms on opposite sides of the equation. We'll gather the [A] terms on one side and the time and reaction rate constant on the opposite side.

$$\text{Equation 5.12} \quad \frac{d[A]}{dt} = -k[A]$$

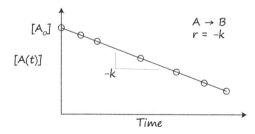

FIGURE 5.2 Plot of concentration versus time for a zero-order reaction.

After gathering like terms on opposite sides of the equation, we have an integral of the general form given in Equation 5.13:

$$\text{Equation 5.13} \quad \int_{[u_0]}^{[u]} \frac{d[u]}{[u]} = \ln(u) - \ln(u_0)$$

where $u = u_0$ when $t = 0$.

When we apply the general form of the solution to Equation 5.12, [A] is given in Equation 5.14.

$$\text{Equation 5.14} \quad \ln([A]) - \ln([A_o]) = \ln\left(\frac{[A]}{[A_o]}\right) = -kt$$

In this case, we can find [A] by taking the exponential of both sides and solving for [A] (Equation 5.15).

$$\text{Equation 5.15} \quad [A] = [A_o]e^{-kt}$$

A linear plot of these data looks much different than the linear plot for zero-order data. Consider data from a test of the first-order degradation of DDT (dichlorodi-phenyltrichloroethane) (Figure 5.3). On a plot of [A] versus time, data that follow first-order reaction kinetics won't be linear; rather, they will exhibit "exponential decay" behavior (Figure 5.3, top). To test data for first-order kinetic behavior, we can either fit the data to the nonlinear form in Equation 5.15 or plot the data in a linearized form. If we reconsider the way we developed the solution for [A] for first-order reaction kinetics, we actually have a linearized form of our data (Figure 5.3). In this case, if a plot of $\ln(A/A_0)$ versus time is linear, then our data follow first-order reaction kinetics, as in Figure 5.3. The reaction rate constant, k, is equal to the slope of the line, and the y-intercept is zero.

Example: Estimating DDT Concentrations over Five Decades

DDT is an insecticide that was used widely in the US from the mid-1940s through the early 1970s. Following the publication of Silent Spring by Rachel Carson (1962), public attention was drawn to the deleterious effects of DDT in particular—and organochlorine compounds in general—in the environment. In 1972, DDT was banned from further use by the US Environmental Protection Agency (1975) after being linked to a wide range of adverse effects on environmental receptors in addition to being identified as a suspected carcinogen through its behavior as an "estrogen mimic." One of the key properties of DDT that made it such an effective insecticide was its persistence in the environment; DDT is especially resistant to destruction by light and natural oxidation processes. The persistence of DDT in the environment is difficult to quantify since site-specific factors and multiple degradation pathways can have a great deal of influence on its degradation. For example, the half-life of DDT in aquatic environments is estimated to be around 150 years (NIH 1998). In contrast, the half-life for DDT in soils is reported to range from 2 to 15 years (DHHS 1994).

Consider a site that was contaminated with DDT from the early 1950s through 1972. In 1975, samples of rhizosphere soils were collected and analyzed to

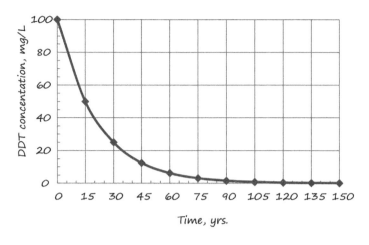

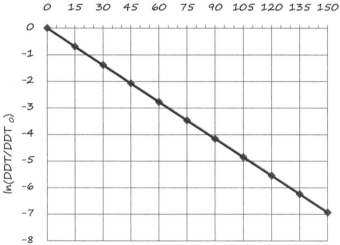

FIGURE 5.3   Plots of DDT concentrations versus time.

determine the level of DDT contamination. The concentration of DDT extracted from the soil was found to be 100 mg/kg. Assume aerobic degradation by rhizosphere bacteria is the only mechanism to reduce DDT concentrations and the reaction is described by first-order reaction kinetics. Estimate the concentration of DDT at the site in 2020.

We're told that the half-life of DDT is 150 yrs. This means that in 150 yrs, $[DDT]/[DDT_0] = 0.5$. We can use this fact to determine the reaction rate constant for DDT degradation by aerobic rhizosphere bacteria by solving Equation 5.16 for k.

$$\text{Equation 5.16}\quad k = \frac{ln(0.5)}{150yr} = 4.62\times10^{-3}/yr$$

Now that we know the reaction rate constant, we can use Equation 5.17 to determine the DDT concentration after 45 years.

$$\text{Equation 5.17} \quad DDT = \left(100\frac{mg}{kg}\right) e^{\left(-\frac{4.62x10^{-3}}{yr}\right)(45\,yr)} = 81.2\frac{mg}{kg}$$

5.3   FACTORS THAT AFFECT THE RATE OF REACTION

Knowledge of the reaction order can tell us a lot about the behavior of a reacting chemical system. But first, we need to examine the factors that affect the reaction rate. Let's begin by assuming that atoms and molecules have to collide with one another in order to react. However, the mere act of colliding isn't sufficient to initiate a reaction. Rather, sufficient energy must be imparted on the colliding atoms/molecules to meet or exceed the thermodynamic threshold necessary for the reaction to occur. This is known as the activation energy, E_{act}. A useful tool to visualize this process is a plot of chemical potential energy versus the reaction coordinate (see Figure 5.4). From these plots, we're able to track changes in energy from the beginning to the end of a reaction. We can also determine if a reaction is exo- or endothermic and calculate the relative rate of a reaction. For example, in the reaction, $A \rightarrow B$, reactant A is transformed into product B, which is at a lower net chemical potential energy. For this reaction to proceed, a certain amount of energy, E_{act}, is needed to make the reaction occur. When comparing reactions with different or multiple reaction pathways, the speed of the reaction is inversely proportional to magnitude of E_{act}. The net change in chemical potential energy is given by ΔE. When $\Delta E < 0$, the reaction is exothermic (heat is given off) as in Figure 5.4. If $\Delta E > 0$, the reaction is endothermic.

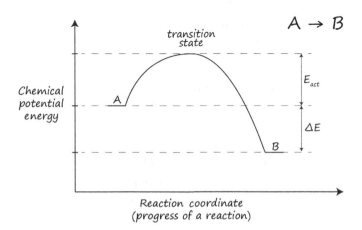

FIGURE 5.4   Plot of chemical potential energy versus reaction coordinate for the reaction $A \rightarrow B$.

So, any factor that can increase the likelihood of collisions between atoms/molecules will increase the reaction rate. For example, increasing the number reactants in a vessel with fixed dimensions can increase the reaction rate. Likewise, increasing the temperature will enhance the mobility of atoms/molecules; consequently, the reaction rate will increase.

Consider a schematic reaction for the aerobic degradation of organic material, OM, presented in Equation 5.18.

Equation 5.18 $OM + O_2 + nutrients \rightarrow H_2O + CO_2 + cells + energy + byproducts$

In this reaction, organic matter is oxidized by aerobic microorganisms while consuming dissolved oxygen and nutrients. We can write a general expression for the schematic reaction rate (Equation 5.19).

Equation 5.19 $r = -k[OM]^w [nutrients]^x [O_2]^y$

You might be wondering why the concentrations of products aren't included in Equation 5.19. In most environmental systems, reaction rates are independent of product concentrations. If we planned to use the reaction in Equation 5.18 to describe the metabolism of organic matter in water, as in the case of activated sludge treatment of domestic wastewater, we would supply sufficient dissolved oxygen and nutrients to the system. As a result, r wouldn't be dependent on [nutrients] or $[O_2]$, and Equation 5.19 could be written in terms of the rate-limiting reactant, [OM] (Equation 5.20).

Equation 5.20 $r = -k[OM]^w$

In some natural and engineered systems, the dissolved oxygen concentration can also be rate limiting. Reaction kinetics for this case would be described using Equation 5.21.

Equation 5.21 $r = -k[OM]^w [O_2]^y$

If we extend this thinking to natural systems, where reactants aren't added under controlled conditions, the availability of nutrients might also become rate limiting and a model similar to Equation 5.19 would apply.

Based on a comparison of Equation 5.19, Equation 5.20, and Equation 5.21, the mathematics needed to describe reaction kinetics for multiple rate-limiting factors can be substantially more involved than the case of a single rate-limiting reactant.

5.4   REACTION CONSTANTS AT NONSTANDARD TEMPERATURES

In natural systems, calculations performed at standard environmental conditions (25°C and 1 atm) often only provide a starting point as we work to describe or predict outcomes of chemical processes. In some cases, it's necessary to know the rate of a reaction at nonstandard temperatures. For example, it's often necessary to examine process behavior at high and low temperatures to predict seasonal behavior, to determine temperature-based rate-limiting behavior, *etc.* In this case, the two-temperature

Arrhenius equation (Equation 5.22) can be used to adjust a known reaction constant at a known temperature to an unknown temperature.

$$\text{Equation 5.22} \quad \frac{k_{T_1}}{k_{T_2}} = e^{\left(\frac{E_{act}}{R}\right)\left(\frac{T_1 - T_2}{T_1 T_2}\right)}$$

where k_{T1} = reaction rate constant at T_1; k_{T2} = reaction rate constant at T_2; T_1 = temperature 1, K; T_2 = temperature 2, K; E_{act} = activation energy, J/mol; R = universal gas constant, 8.314 J/mol·K. When using Equation 5.22, temperatures should be specified in degrees Kelvin.

In some cases, the activation energy is known through experimentation. For example, E_{act} ranges from 33,500 to ~50,000 J/mol for microbial systems. While E_{act} might not be known in all instances, in most environmental applications, Equation 5.23 can be written as:

$$\text{Equation 5.23} \quad \frac{k_{T_1}}{k_{T_2}} = \theta^{(T_1 - T_2)}$$

In Equation 5.23, temperatures are expressed in degrees Celsius. At temperatures between zero and 35 °C, θ = 1.047. Other investigators have recommended different values of θ based on subsets of the previous temperature range; these include θ = 1.135 for temperatures between 4 and 20 °C and θ = 1.056 for T > 20 °C (Flegal and Schroeder 1976; Ingraham 1959).

5.5   ACID-BASE REACTIONS

5.5.1   CHARACTERISTICS OF ACID-BASE REACTIONS

In aquatic systems, dilute solutions of acids (HA) and their conjugate bases (A⁻) (Equation 5.24) are related through the acid equilibrium constant, K_A, as in Equation 5.25.

$$\text{Equation 5.24} \quad HA + H_2O \overset{K_A}{\leftrightarrow} A^- + H_3O^+$$

$$\text{Equation 5.25} \quad K_A = \frac{\left[A^-\right]\left[H_3O^+\right]}{\left[HA\right]} = \frac{\left[A^-\right]\left[H^+\right]}{\left[HA\right]}$$

In aqueous solutions, "free protons," H^+, actually exists as protonated water molecules, H_3O^+. However, from a practical point of view, we can treat them as protons. The total concentration of species "A" in the system at any time is (Equation 5.26):

$$\text{Equation 5.26} \quad C_A = \left[HA\right] + \left[A^-\right]$$

This type of reaction is known as a Bronsted-Lowry acid-base reaction, where acids are species that donate protons and bases are species that accept protons. K_A values for a variety of acids are presented in Table 5.1. When K_A is large, the products side of Equation 5.24 is favored. In this case, protons are stripped from the acid, HA, and the conjugate base dominates. Conversely, when K_A is low, the protonated form of the

TABLE 5.1
Acid Equilibrium Constants K_A and pK_A for Common Acids (at 25 °C)
(Haynes 2014).

Acid	Equilibrium Equation	K_A	pK_A
Hydrochloric acid	$HCl \rightarrow H^+ + Cl^-$	~10^3	~ −3
Sulfuric acid	$H_2SO_4 \rightarrow 2H^+ + SO_4^{2-}$	~10^3	~ −3
Phosphoric acid	$H_3PO_4 \leftrightarrow H^+ + H_2PO_4^-$	$10^{-2.1}$	2.1
Carbonic acid and dissolved carbon dioxide	$H_2CO_3^* \leftrightarrow H^+ + HCO_3^-$	$10^{-6.3}$	6.3
Dihydrogen phosphate ion	$H_2PO_4^- \leftrightarrow H^+ + HPO_4^{2-}$	$10^{-7.2}$	7.2
Hypochlorous acid	$HOCl \leftrightarrow H^+ + OCl^-$	$10^{-7.5}$	7.5
Ammonium ion	$NH_4^+ \leftrightarrow H^+ + NH_3$	$10^{-9.3}$	9.3
Bicarbonate ion	$HCO_3^- \leftrightarrow H^+ + CO_3^{2-}$	$10^{-10.3}$	10.3
Monohydrogen phosphate ion	$HPO_4^{2-} \leftrightarrow H^+ + PO_4^{3-}$	$10^{-12.3}$	12.3
Water	$H_2O \leftrightarrow H^+ + OH^-$	10^{-14}	14

acid is favored. Since values of K_A are spread over many orders of magnitude, it can be convenient to consider $-\log_{10}(K_A)$ or "pK_A" as opposed to K_A. This approach can also make it much easier to develop graphical representations of acid-base equilibria.

The strength of an acid is related directly to the tendency of an acid to donate a proton(s). In particular, strong acids dissociate rapidly and completely in water. As a general guideline, strong acids have $K_A > 1$ ($pK_A < 1$). In Table 5.1, hydrochloric acid and sulfuric acids are strong acids while the others are weak acids. You'll note that the reactions for strong acids are written to favor the products side of the reaction. When added to water, we assume strong acids dissociate immediately into their constituent ions.

Acids are also characterized by the number of protons they can accept or donate. For example, the acid in Equation 5.24 donates only one proton and is known as a monoprotic acid. In contrast, carbonic acid is a diprotic acid and its equilibrium is described by two reactions, each with its own K_A (Equation 5.27 and Equation 5.28).

$$\text{Equation 5.27} \quad H_2CO_3 \leftrightarrow H^+ + HCO_3^-$$

$$\text{Equation 5.28} \quad HCO_3^- \leftrightarrow H^+ + CO_3^{2-}$$

Similarly, triprotic acids can exchange three protons. In Table 5.1, phosphoric acid is an example of a triprotic acid.

5.6   SELF-IONIZATION OF WATER AND PH

Water can act as both an acid and a base. In an aqueous solution, a water molecule deprotonates to form an OH^- ion and a hydrogen nucleus, H^+. The proton that's liberated from the water molecule then protonates another water molecule to form a hydronium ion, H_3O^+. The equilibrium equation that describes these processes is shown in Equation 5.29 with an equilibrium reaction constant, K_w, given in Equation 5.29.

$$\text{Equation 5.29} \quad H_2O + H_2O \overset{K_w}{\longleftrightarrow} H_3O^+ + OH^-$$

$$\text{Equation 5.30} \quad K_W = [H_3O^+][OH^-] = [H^+][OH^-] = 10^{-14} \text{ at } 25^\circ C$$

You might wonder why the two water molecules in Equation 5.29 don't appear in the denominator of Equation 5.30. In this case, water is the continuous phase, so we can assume it's present in significant excess—essentially a constant concentration relative to other reacting chemical species. Consequently, the concentration of water molecules is actually incorporated into the constant, K_w. Taking the $\log_{10}$ of both sides of Equation 5.31 yields the following:

$$\text{Equation 5.31} \quad log\left[H^+\right] + log\left[OH^-\right] = -14$$

In chemistry, "p" is used as a shorthand notation for the negative log of a concentration. As a result, we can write pH and pOH according to Equation 5.34, Equation 5.32, and Equation 5.33, respectively.

$$\text{Equation 5.32} \quad pH = -log\left[H^+\right]$$

$$\text{Equation 5.33} \quad pOH = -log\left[OH^-\right]$$

These relationships allow us to write Equation 5.31 in terms that might be more familiar (Equation 5.34).

$$\text{Equation 5.34} \quad pH + pOH = 14$$

When pH is neutral, pH = 7 and pOH = 7. For a solution with a pH of 6, the pOH is 8.

5.7 CLASSIFICATION OF THERMODYNAMIC SYSTEMS

Thermodynamic systems are defined based on the transfer of matter and energy, as shown in Figure 5.5. In open systems, both mass and energy can be exchanged with the surrounding environment, while only energy is exchanged in closed systems. In isolated systems, neither mass nor energy is exchanged. Experimentally, it's common to approximate an isolated system by enclosing a closed vessel in multiple layers of thermal insulation. It should come as no surprise that isolated systems are much less relevant than open and closed systems when studying natural water systems.

While the preceding observations may seem straightforward, the differences between these systems are actually quite profound and can have major impacts on the characteristics of aquatic systems. If the Earth is treated as a closed system, matter isn't exchanged across the system boundary with space, but energy can be exchanged. In this regard, the total amount of water on Earth is finite. As an additional caveat, it is assumed that the only "pure" water is vapor-phase water. As vapors condense, gases in the atmosphere dissolve in liquid water. Similarly, as liquid water moves through the environment, it can acquire a variety of soluble constituents, including minerals, nutrients, and a variety of natural and manmade organic compounds.

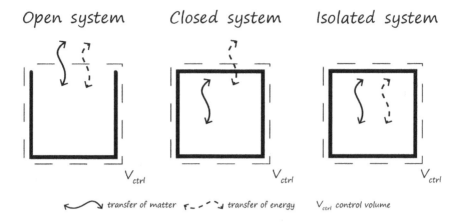

Open system   Closed system   Isolated system

↝ transfer of matter   ⤏ transfer of energy V_{ctrl} control volume

FIGURE 5.5 Schematic representation of open, closed, and isolated thermodynamic systems.

5.8 CARBONATE EQUILIBRIUM IN NATURAL WATERS

The significance of having a water system open to the atmosphere might not be immediately apparent. While the atmosphere contains many gases that are soluble in water, the roles of oxygen and inorganic carbon gases on natural waters are of paramount importance. Let's focus on an open water sample in equilibrium with atmospheric CO_2 (g). Let's also assume there are no other sources of carbonate in this system (*e.g.*, calcium carbonate solids).

At the air-water interface, carbon dioxide is transferred between the aqueous and gas phases as represented in Equation 5.35.

$$\text{Equation 5.35} \quad CO_2(g) \leftrightarrow CO_2(aq)$$

The equilibrium relationship between a gas and its aqueous phase concentration is given by Henry's law (Henry 1803) (Equation 5.36).

$$\text{Equation 5.36} \quad \left[C(aq)\right] = K_H p_g$$

where $[C_{aq}]$ is the aqueous phase concentration of gas g, K_H is the Henry's law constant for gas g, and p_g, is the partial pressure of the gas. Values of K_H for a variety of environmentally relevant gases are presented in Table 5.2. Two reliable sources of thermodynamic data, including Henry's law constants, include the CRC Handbook of Chemistry and Physics Online (https://hbcp.chemnetbase.com/faces/contents/ContentsSearch.xhtml) and the National Institute of Standards and Technology's Chemistry WebBook (http://webbook.nist.gov/chemistry). Using this information, we can determine the aqueous phase concentration of CO_2 in water that's open to the atmosphere (Equation 5.37) based on its partial pressure ($10^{-3.5}$ atm) and the Henry's law constant ($10^{-1.5}$ mol/L·atm at 25 °C).

$$\text{Equation 5.37} \quad \left[CO_2(aq)\right] = \left(10^{-3.5}\,atm\right)\left(10^{-1.5}\,\frac{mol}{L\,atm}\right) = 10^{-5}\,M$$

TABLE 5.2

Henry's Law Equilibrium Reactions and Constants for Select Gases at 25 °C (Haynes 2014).

Reaction	K_H, mol L^{-1} atm^{-1}
Cl_2 (g) $\leftrightarrow$ Cl_2 (aq)	$10^{-1.03}$
CO (g) $\leftrightarrow$ CO (aq)	$10^{-3.51}$
CO_2 (g) $\leftrightarrow$ CO_2 (aq)	$10^{-1.48}$
NH_3 (g) $\leftrightarrow$ NH_3 (aq)	$10^{1.78}$
NO_2 (g) $\leftrightarrow$ NO_2 (aq)	$10^{-1.96}$
O_2 (g) $\leftrightarrow$ O_2 (aq)	$10^{-2.92}$
O_3 (g) $\leftrightarrow$ O_3 (aq)	$10^{-3.92}$

Henry's law works well when describing the solubility of gases in water at low concentrations and low partial pressures. These conditions are often known as "ideal dilute solutions." Experimentally, the solubility of CO_2(g) in water decreases with increasing temperature.

Aqueous phase carbon dioxide interacts with water molecules to form carbonic acid, H_2CO_3 (aq) as described in Equation 5.38 and Equation 5.39.

$$\text{Equation 5.38} \quad CO_2\left(aq\right) + H_2O \leftrightarrow H_2CO_3\left(aq\right)$$

$$\text{Equation 5.39} \quad K = \frac{\left[H_2CO_3\left(aq\right)\right]}{\left[CO_2\left(aq\right)\right]} = 10^{-2.8}$$

Both CO_2 (aq) and H_2CO_3 (aq) are acids that predominate at pH $\ll$ 6.3. Analytically, it's difficult to distinguish between the concentrations of dissolved carbon dioxide and carbonic acid in solution. To simplify calculations, it's convenient to treat the sum of the aqueous phase carbon dioxide and carbonic acid concentrations as one entity, known as H_2CO_3 (aq)* (Equation 5.40).

$$\text{Equation 5.40} \quad \left[H_2CO_3\left(aq\right)^*\right] = \left[CO_2\left(aq\right)\right] + \left[H_2CO_3\left(aq\right)\right]$$

Based on the magnitude of the equilibrium constant in Equation 5.39, the carbonic acid concentration is much less than [CO_2 (aq)]. This allows us to simplify Equation 5.40 as shown in Equation 5.41.

$$\text{Equation 5.41} \quad \left[H_2CO_3\left(aq\right)^*\right] \cong \left[CO_2\left(aq\right)\right]$$

Now that we've developed an idea of how CO_2 (g) partitions between the gas and aqueous phases to interact with water and form carbonic acid, we need to consider the acid/base behavior of H_2CO_3 (aq)* in water. Since we know that carbonate is a diprotic acid, we know that three carbonate species will form, depending on pH. The equilibria for these carbonate species are given in Equation 5.42 through Equation 5.45.

$$\text{Equation 5.42} \quad H_2CO_3(aq)^* \overset{K_{A_1}}{\leftrightarrow} HCO_3^-(aq) + H^+$$

$$\text{Equation 5.43} \quad K_{A_1} = \frac{\left[HCO_3^-(aq)\right]\left[H^+\right]}{\left[H_2CO_3(aq)^*\right]} = 10^{-6.3}$$

$$\text{Equation 5.44} \quad HCO_3^-(aq) \overset{K_{A_2}}{\leftrightarrow} H^+ + CO_3^{2-}(aq)$$

$$\text{Equation 5.45} \quad K_{A_2} = \frac{\left[CO_3^{2-}(aq)\right]\left[H^+\right]}{\left[HCO_3^-(aq)\right]} = 10^{-10.3}$$

To fully describe the system, we need to develop a mass balance expression for the total concentration of carbonate, C_T (Equation 5.46).

$$\text{Equation 5.46} \quad C_T = \left[H_2CO_3(aq)^*\right] + \left[HCO_3^-(aq)\right] + \left[CO_3^{2-}(aq)\right]$$

For a known pH, the concentrations of H_2CO_3 (aq)*, HCO_3^- (aq), and CO_3^{2-} (aq) can be determined by solving the system of Equation 5.46, Equation 5.45, Equation 5.43, and Equation 5.41 simultaneously. If we linearize these equations and plot their $-\log_{10}[C]$ versus pH, we can develop a graphical representation of carbonate species as a function of pH in an open system (Figure 5.6). The procedure to construct these diagrams is beyond the scope of this book. However, speciation diagrams can be useful tools for understanding the relationships between a variety of aqueous phase chemical constituents.

So, what's the practical significance of having a water system open to carbon dioxide in the atmosphere? When an open water system is open to the atmosphere, CO_2 (g) serves as a constant source of 10^{-5} M of carbonate. As pH is increased above ~6.3, bicarbonate and carbonate species are formed, and the total concentration of carbonate increases proportionally. Since CO_2 (aq) is a weak acid, natural water in equilibrium with atmospheric CO_2 will have a pH <7.

Now, let's consider a closed carbonate system. In this case, atmospheric carbon dioxide doesn't serve as a reservoir of carbon dioxide so we need to specify a total concentration of carbonate. Let's assume $C_T = 10^{-5}$ M. In this system, the relative concentrations of carbonate species change as a function of pH. We can linearize the equilibria and mass balance expressions to develop a graphical representation of carbonate speciation in a closed system, as in Figure 5.7. Treating the closed carbonate system as a diprotic acid, we know that $H_2CO_3^*$ will be the predominant carbonate species at pH < 6.3 (pK_{A1}). Similarly, HCO_3^- will predominate at pH values between pK_{A1} and pK_{A2}. At pH > pK_{A2}, the majority of carbonate will be present as CO_3^{2-}.

Now, we can compare carbonate equilibrium in open and closed systems. In the open system, atmospheric CO_2 serves as a baseline source of carbonate that supports a 10^{-5} M minimum total carbonate concentration. At pH > pK_{A1}, the total carbonate concentration becomes significantly greater than 10^{-5} M in the open system. In the closed system, however, the total concentration of carbonate won't exceed 10^{-5} M.

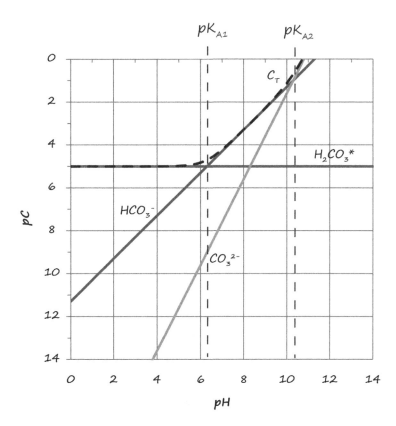

FIGURE 5.6 Plot of pC versus pH for a water system open to the atmosphere assuming standard environmental conditions, a dilute solution, and the absence of other carbonate solids.

5.9   SOLUBILITY AND THE SOLUBILITY PRODUCT

Solubility, S, is the amount of a solute that can be dissolved in a solvent. Before we proceed, it's important to recognize that the solid has to be present if we're going to use solubility relationships to estimate the equilibrium composition of solutions. Likewise, these relationships apply to saturated solutions (*i.e.*, solutions in which no additional solute will dissolve). The equilibrium reaction rate constant used to describe the relationship between a precipitate (solid) and its constituent ions is known as the solubility product, K_{sp}.

Consider the dissolution of aluminum hydroxide shown in Equation 5.47. At equilibrium, this reaction is described by the solubility product (Equation 5.48).

$$\text{Equation 5.47}\quad Al\left(OH^-\right)_3(s) \overset{K_{sp}}{\longleftrightarrow} Al^{3+} + 3OH^-$$

$$\text{Equation 5.48}\quad K_{sp} = \left[Al^{3+}\right]\left[OH^-\right]^3$$

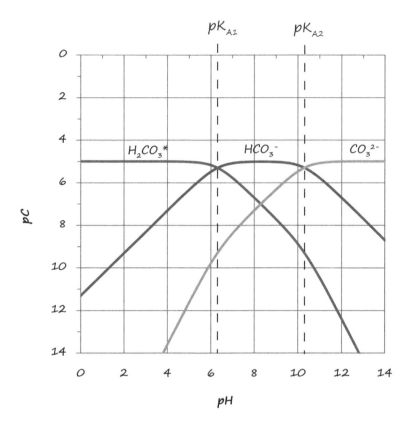

FIGURE 5.7   pC-pH diagram for a water system that's closed to the atmosphere assuming $C_T = 10^{-5}$ M, standard environmental conditions, a dilute solution, and the absence of other sources of carbonate.

Solubility products for common reactions of significance in water systems are presented in Table 5.3. In general, the solubility of metals decreases as the solvent (water) pH is increased. In the case of $Al(OH)_3$ (s), an increase in pH results in a shift in equilibrium that favors the formation of the solid. Similarly, the solubility of salts typically increases with temperature. However, there are a number of significant cases where solubility decreases as temperature is increased. For example, the solubility of limestone, $CaCO_3$ (s), decreases as temperature is raised. This can lead to significant operational and maintenance issues for industries that use hard water in heat exchange processes (boilers, evaporators, condensers, *etc.*)

It's often useful to represent the conditions under which a precipitate forms on a pC-pH diagram (see Figure 5.8). At "A," there isn't enough soluble aluminum for the solid to form, regardless of the solution pH. At "B," there's sufficient soluble aluminum in solution, but the pH is too low for the solid to form. However, if the solution pH is increased to ~5, $Al(OH)_3(\downarrow)$ will precipitate. Note that at pH ~8, the concentration at "B" is again insufficient for aluminum to precipitate as a hydroxide.

TABLE 5.3

Solubility Products for Select Reactions (at Standard Conditions) (Haynes 2014).

Reaction	K_{sp}
$Al(OH)_3(s) \rightarrow Al^{3+} + 3OH^-$	$10^{-31.2}$
$CaCO_3(s) \rightarrow Ca^{2+} + CO_3^{2-}$	$10^{-8.4}$
$Ca(OH)_2(s) \rightarrow Ca^{2+} + 2OH^-$	$10^{-5.4}$
$CaSO_4(s) \rightarrow Ca^{2+} + SO_4^{2-}$	$10^{-4.6}$
$Cd(OH)_2(s) \rightarrow Cd^{2+} + 2OH^-$	$10^{-13.6}$
$FeCO_3(s) \rightarrow Fe^{2+} + CO_3^{2-}$	$10^{-10.4}$
$Fe(OH)_2(s) \rightarrow Fe^{2+} + 2OH^-$	$10^{-14.5}$
$Fe(OH)_3(s) \rightarrow Fe^{3+} + 3OH^-$	$10^{-38.0}$
$MgCO_3(s) \rightarrow Mg^{2+} + CO_3^{2-}$	$10^{-4.9}$
$Mg(OH)_2(s) \rightarrow Mg^{2+} + 2OH^-$	$10^{-9.2}$
$Zn(OH)_2(s) \rightarrow Zn^{2+} + 2OH^-$	$10^{-17.1}$

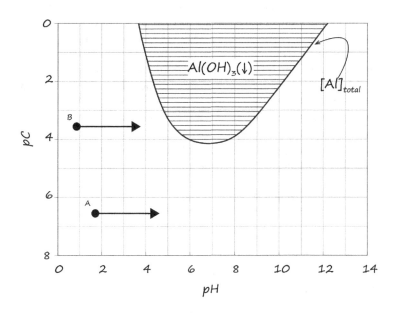

FIGURE 5.8 Solubility diagram for $Al(OH)_3(s)$ where pC = −log [soluble Al species].

5.9.1 COMMON ION EFFECT

Now, consider what happens if we add another compound that shares an ion in common with a saturated salt solution. For instance, how would the aluminum equilibrium be affected if aluminum nitrate $(Al(NO_3)_3$ (aq)) was added to a saturated $Al(OH)_3$ solution? In this case, by adding a common ion (Al^{3+}) to the saturated solution, equilibrium shifts and aluminum precipitates. The addition of a common ion decreases solubility.

5.9.2 COMPLEXATION REACTIONS

Metal complexes (or "coordinated compounds") are soluble species that are formed when a central metal ion is covalently bonded to an anion known as a ligand. A schematic reaction for the formation of a complex ion is presented in Equation 5.49.

$$\text{Equation 5.49} \quad Me^{n+} + ligand \overset{K_s}{\leftrightarrow} Me-ligand$$

where K_s = equilibrium stability constant. More stable complexes have higher stability constants.

Examples of inorganic and organic molecules that can serve as ligands include phosphate, carbonate, chloride, cyanide, hydroxide, nitrate, ethylenediaminetetraacetic acid (EDTA), and nitrilotriacetic acid (NTA). Like many areas of science, it's difficult to provide hard and fast rules to determine the strength or stability of complexes. This is due in large part to the number of exceptions to any rule. However, a few generalities can be made. "A metals" such as Ca^{2+}, Mg^{2+}, Al^{3+}, and Si^{4+} coordinate with ligands that contain oxygen as the electron donor (*e.g.*, CO_3^{2+}, OH^-, and BO_3^{3-}). "B metals" such as Ag^+, Cu^+, and Ni^{2+} tend to form complexes with ligands that contain N, S, and P. Transition metals, including chromium and iron, are able to form complexes with many different ligands. Additionally, complex stability is also higher for central metal ions that have a higher valence charge (Stumm and Morgan 1996; Snoeyink and Jenkins 1991).

Hydrolysis reactions—the interaction of a central metal ion with hydroxide as a ligand—are a specific type of complexation reaction that is especially significant to freshwater aquatic systems. A general form of the hydrolysis reaction and the corresponding stability constant are given in Equation 5.50 and Equation 5.51, respectively.

$$\text{Equation 5.50} \quad Me^{n+} + OH^- \overset{K_s}{\leftrightarrow} Me-ligand^{n-1}$$

$$\text{Equation 5.51} \quad K_s = \frac{\left[Me-ligand^{n-1}\right]}{\left[Me^{n+}\right]\left[OH^-\right]}$$

These relationships can be related to the proton concentration using the self-ionization of water (Equation 5.52).

$$\text{Equation 5.52} \quad K_s = \frac{\left[Me-ligand^{n-1}\right]\left[H^+\right]}{\left[Me^{n+}\right]\left[K_w\right]}$$

To better understand hydrolysis and to learn how the formation of metal complexes can impact metal solubility, let's consider a water solution to which a zinc salt is added. The reaction that describes the solubility of zinc in water is shown in Equation 5.53.

$$\text{Equation 5.53} \quad Zn(OH)_2(s) \xleftrightarrow{10^{-17.1}} Zn^{2+} + 2OH^-$$

For a solution of zinc in water, the equilibrium stability reactions for the four hydroxide complexes that form are given in Equation 5.54 through Equation 5.57.

$$\text{Equation 5.54} \quad Zn^{2+} + OH^{-} \xleftrightarrow{K_{s1}=10^{4.2}} ZnOH^{+}(aq)$$

$$\text{Equation 5.55} \quad ZnOH^{+}(aq) + OH^{-} \xleftrightarrow{K_{s2}=10^{6}} Zn(OH)_{2}^{0}(aq)$$

$$\text{Equation 5.56} \quad Zn(OH)_{2}^{0}(aq) + OH^{-} \xleftrightarrow{K_{s3}=10^{4.1}} Zn(OH)_{3}^{-}(aq)$$

$$\text{Equation 5.57} \quad Zn(OH)_{3}^{-}(aq) + OH^{-} \xleftrightarrow{K_{s4}=180} Zn(OH)_{4}^{2-}(aq)$$

The total soluble concentration of Zn in the solution is given in Equation 5.58.

$$\text{Equation 5.58} \quad C_{T} = \left[Zn^{2+}\right] + \left[ZnOH^{+}(aq)\right] + \left[Zn(OH)_{2}^{0}(aq)\right]$$
$$+ \left[Zn(OH)_{3}^{-}(aq)\right] + \left[Zn(OH)_{4}^{2-}(aq)\right]$$

Before proceeding further, readers might wonder, "how we can know exactly what central metal ion-ligand complexes will form?" The only way to know if a reaction is possible is to determine the Gibbs' free energy of a prospective reaction. Stability constants for select environmentally relevant metal hydroxide complexes are presented in Table 5.4.

While it can be challenging to see how we can use the equilibrium reactions for metal-ligand complexes in a direct way, they actually form the basis for graphical representations of metal solubility that are much more easily applied. Just as we saw in the section on acid-base reactions, developing the graphical representations of these equilibria is outside the scope of this book. However, the application of speciation (pC-pH) diagrams is not. The development of pC-pH diagrams can be a useful problem-solving strategy that incorporates the equilibrium associated with the self-ionization of water, the formation of a metal precipitate, and the predominance of complex ions as a function of pH. For the case of zinc in water described earlier, the

TABLE 5.4
Stability Constants for Metal Hydroxide Complexes (Haynes 2014).

Central Metal Ion	$\log_{10}K_{s1}$	$\log_{10}K_{s2}$	$\log_{10}K_{s3}$	$\log_{10}pK_{s4}$
Al^{3+}	9	17.7	25.3	33.3
Fe^{2+}	4.6	7.4	11	10
Fe^{3+}	11.81	22.4	30.2	34.4
Pb^{2+}	6.4	10.9	13.9	–
Zn^{2+}	5	11.1	13.6	14.8

$K_{s1} = ([\text{Me-OH}^{n-1}])/[\text{Me}^{n+}][\text{OH}^{-}]$, $K_{s2} = [\text{Me-OH}^{n-2}]/[\text{Me-OH}^{n-1}][\text{OH}^{-}]$, $K_{s3} = [\text{Me-OH}^{n-3}]/[\text{Me-OH}^{n-2}][\text{OH}^{-}]$, $K_{s4} = [\text{Me-OH}^{n-4}]/[\text{Me-OH}^{n-3}][\text{OH}^{-}]^{4}$

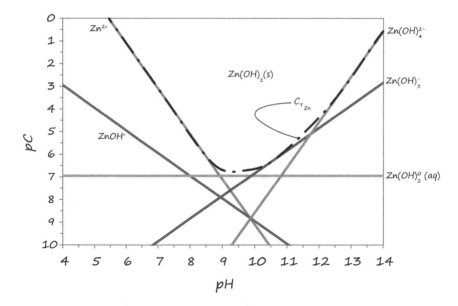

FIGURE 5.9 pC-pH diagram for soluble hydrated zinc complexes assuming standard environmental conditions, a dilute solution, and the absence of any other sources of zinc.

equilibrium constants can be expressed as a function of the concentration of relevant zinc species and pH. By taking the log of both sides of the equilibrium reaction constants and the mass balance equation, we can develop and linearize expressions that can be plotted to establish the predominance of Zn species as well as Zn solubility as a function of pH. For example, the linearized forms of soluble hydrated zinc complexes are shown graphically in Figure 5.9.

Graphic representations of metal solubilities can provide a tremendous amount of useful information. For example, using the data in Figure 5.9, we can see that Zn has its minimum solubility in water at a pH of ~9.3. Likewise, if we had a solution containing a total zinc concentration of 10^{-3} M, we can see that the pH of the water would have to be greater than ~6.9 for zinc to precipitate as a hydroxide. Similarly, if we wanted to reduce the zinc concentration to less than 10^{-5} M in a water solution, the pH would have to range from 8 to a little less than 12. It's important to remember that while predominance diagrams can be useful, full calculations of the system of equilibrium equations should be performed when determining actual system points.

Since complexation reactions involve free metal ions, we need to consider their subsequent impact on metal solubility. For instance, the concentration of soluble zinc species increases when Zn ions are removed from the solubility equilibria as Zn-OH complexes are formed. For example, if another ligand such as phosphate is added, soluble $ZnHPO_4^0$ (aq) is formed through the reaction of Zn^{2+} with HPO_4^{2-} and PO_4^{3-} (Sjöberg *et al.* 2013). In this case, the concentration of $ZnHPO_4^0$ (aq) has to be added to the mass balance equation (Equation 5.59).

Equation 5.59
$$C_T = \left[Zn^{2+}\right] + \left[ZnOH^+(aq)\right] + \left[Zn(OH)_2^0(aq)\right]$$
$$+ \left[Zn(OH)_3^-(aq)\right] + \left[Zn(OH)_4^{2-}(aq)\right] + \left[Zn(HPO)_4^0(aq)\right]$$

The solubility of zinc will increase further if additional ligands are added to solution.

Example: Using Predominance Diagrams to Estimate the Chemical Composition of a Stream

In northern Minnesota's iron range, a treated mine drainage stream is discharged into a well oxygenated freshwater stream. The pH of the stream is circumneutral and contains sufficient alkalinity to resist changes in pH. Use the pC-pH predominance diagram(s) for the solubility of ferrous and ferric iron hydroxides (Figure 5.10 and Figure 5.11, respectively), to determine the total concentration of soluble ferric iron (Fe^{3+}) in solution. Also, estimate the soluble Fe^{2+} concentration. State all assumptions and justify your answers.

Assumptions: The ions are in a dilute solution under standard environmental conditions. No ligands other than OH^- are present. The solids, $Fe(OH)_2$ (s) and

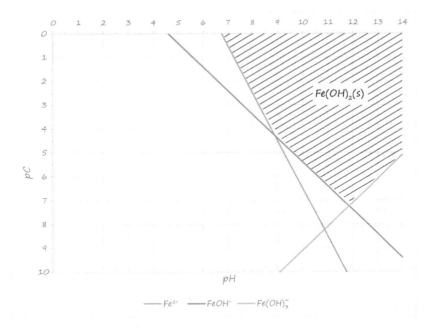

Mass balance on Fe^{2+}:
$[Fe^{3+}_{Total}] = [Fe^{2+}] + [FeOH^+] + [Fe(OH)_3^-]$

Assume a dilute solution of Fe^{2-} in equilibrium with water under standard conditions. Neglect the presence of ligands other than hydroxide.

FIGURE 5.10　Predominance diagram for the hydrolysis of Fe^{2+}.

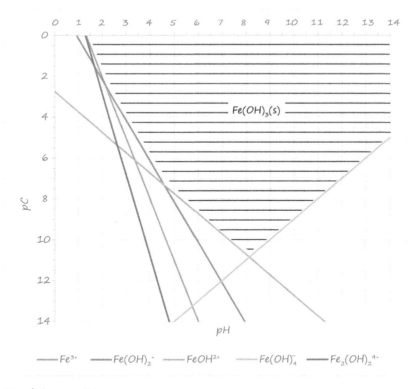

Mass balance on Fe^{3+}:

$$[Fe^{3+}_{Total}] = [Fe^{3+}] + [Fe(OH)_2^+] + [FeOH^{2+}] + [Fe(OH)_4^-] + [Fe_2(OH)_2^{4+}]$$

Assume a dilute solution of Fe^{3+} in equilibrium with water under standard conditions. Neglect the presence of ligands other than hydroxide.

FIGURE 5.11   Predominance diagram for the hydrolysis of Fe^{3+}.

$Fe(OH)_3$ (s) are present. In a well oxygenated stream, Fe^{2+} is oxidized to form Fe^{3+}. Consequently, it's reasonable to assume that the concentration of Fe^{2+} is low when compared with that of Fe^{3+}. The Fe^{3+} concentration at pH = 7 can be found directly from Figure 5.11. In this case, $[Fe^{3+}] \sim 10^{-9.5}$ M. While there was no need to use the pC-pH diagram for ferrous iron in this case, the speciation of ferrous iron can be important in low oxygen waters—including groundwater.

REFERENCES

Burgess, D. (2004). *NIST SRD 46. Critically Selected Stability Constants of Metal Complexes: Version 8.0 for Windows*, National Institute of Standards and Technology, US Department of Commerce, https://doi.org/10.18434/M32154, accessed April 18, 2023.

Carson, R. (1962). *Silent Spring*, Houghton Mifflin Riverside Press, Cambridge, MA, 368 pp.

DHHS (1994). "Toxicology Profile for 4,4' -DDT, 4,4' -DDE, 4,4' -DDD (Update)", Agency for Toxic Substances and Disease Registry, US Department of Human Health and Human Services, Atlanta, GA, 194 pp.

Flegal, T., and E. Schroeder (1976). "Temperature Effects on BOD Stoichiometry and Oxygen Uptake Rate," *Journal of the Water Pollution Control Federation*, 49 (12), 2700.

Haynes, W. (Ed.) (2014). *CRC Handbook of Chemistry and Physics*, 85/e, CRC Press and Taylor and Francis, Boca Raton, FL, 2666 pp.

Henry, W. "III. (1803). "Experiments on the Quantity of Gases Absorbed by Water, at Different Temperatures, and Under Different Pressures," *Philosophical Transactions of the Royal Society of London*, 93, 29–274.

Hutson, S., Barber, N., Kenny, J., Linsey, K., Lumia, D., and M. Maupin (2004). *Estimated Use of Water in the United States in 2000*, US Geological Survey Circular 1268, US Geological Survey, Reston, VA, 48 pp.

Ingraham, J. (1959). "Growth of Psychrophilic Bacteria," *Bacteriology*, 76 (1), 75.

International Union of Pure and Applied Chemistry (IUPAC) (2005). *Stability Constants Database*, International Union of Pure and Applied Chemistry, London.

International Union of Pure and Applied Chemistry-National Institute for Standard and Technology (IUPAC-NIST) (2012). *Solubility Database, Version 1.1, NIST Standard Reference Database 106*, International Union of Pure and Applied Chemistry-National Institute for Standard and Technology Solubility Data Series, http://dx.doi.org/10.18434/T4QC79.

Martell, A., Smith, R., and R. Motekaitis (2004). *NIST Standard Reference Database 46 Version 8.0: NIST Critically Selected Stability Constants of Metal Complexes*, National Institute of Standards and Technology, Gaithersburg, MD.

National Institutes of Health (NIH) (1998). *The Hazardous Substances Data Bank (HSDB) [CD-ROM]; U.S. National Library of Medicine; National Institutes of Health*, US Department of Health and Human Services, Bethesda, MD.

Sander, R. (2001). *In: NIST Chemistry WebBook, NIST Standard Reference Database Number 69*, P. Linstrom and W. Mallard, Eds, National Institute of Standards and Technology, Gaithersburg, MD, doi:10.18434/T4D303.

Sjöberg, S., Powell, K., Brown, P., Byrne, R., Gajda, T., Hefter, G., Leuz, A., and H. Wanner (2013). "Chemical Speciation of Environmentally Significant Metals with Inorganic Ligands. Part 5: The Zn^{2+} + OH^-, Cl^-, CO_3^{2-}, SO_4^{2-}, and PO_4^{3-} systems," *Pure and Applied Chemistry*, 85 (12), 2249–2311.

Snoeyink, V., and D. Jenkins (1991). *Water Chemistry*, John Wiley and Sons, New York, 480 pp.

Stumm, W., and J. Morgan (1996). *Aquatic Chemistry: Chemical Equilibria and Rates in Natural Waters, 3/e*, John Wiley and Sons, New York, 103–105 pp.

US Environmental Protection Agency (USEPA) (1975). *DDT: Review of Scientific and Economic Aspects of the Decision to Ban its Use as a Pesticide, EPA-540/1-75-022; US Environmental Protection Agency, Office of Pesticide Programs*, US Government Printing Office, Washington, DC.

6 Tracking the Movement of Mass in Aquatic Environmental Systems

6.1 MATERIAL BALANCES

Mass (matter) and energy are conserved substances—they can neither be created nor destroyed. As we saw in the previous chapter, a balanced chemical reaction must contain equal masses of products and reactants. Likewise, the charge on the product and reactant sides of a balanced reaction must be equal. Similarly, when a chemical reaction occurs, chemical potential energy is consumed as bonds are broken and formed. Heat—thermal energy—can also be created or consumed. In both cases, the total mass and energy before and after the reaction must be equal. This concept will become the foundation for the ways we analyze the movement and interaction of constituents in and between environmental compartments. To begin this discussion, we're going to develop a quantitative approach to track the rate that materials move in water systems. This approach will enable us to understand and describe a wide range of practical circumstances, including the rate that pollutants accumulate in a system or the concentration of a pollutant in a river after an upstream spill.

We can track the movement of mass into and out of this system in a manner similar to the way you would balance a bank account. Instead of a register of deposits and withdrawals, we'll consider an environmental compartment that has a boundary that we'll call a control volume (Figure 6.1). This could be a beaker in the lab, a river reach, a lake, or the atmosphere. A deposit in your bank account results in an increase in the funds in your account and a withdrawal leads to a decrease in the amount of money. Your account can also "generate" money by earning interest over time. The amount of interest you earn is proportional to the amount of money in your account and the interest rate that your money earns. In our control volume, we use arrows to visualize the different paths through which mass enters and leaves our system. In some cases, mass can be converted to other products through (bio)chemical reactions. This is similar to the interest term in your bank account. By balancing the flow of mass into and out of the system and accounting for the addition or removal of mass through reactions, we can determine the rate that mass is accumulated in the control volume.

In practice, the terms in the general material balance statement are represented in units of mass·time^{-1} since the systems we encounter in environmental applications most typically involve the movement of water and substances that can react over time. This idea is represented in the general word-based material balance statement presented in Figure 6.1.

 DOI: 10.1201/9781003289630-7

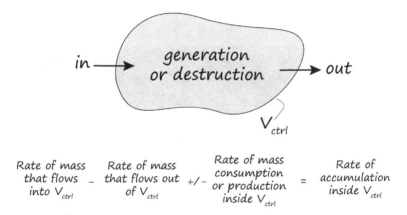

FIGURE 6.1   Control volume and mass or energy flows in an environmental compartment and a word statement for the material balance.

When setting a control volume, its boundaries must be clearly defined. This will help us to identify the inputs and outputs to and from the system, respectively. Additionally, setting a particular control volume should result in practical information. In some cases, the choice of a control volume is evident. For example, if we wanted to study the reaction of two chemicals in a batch experiment, the beaker would serve as an appropriate control volume.

In natural systems, the selection of a control volume, V_c, is determined largely by site-specific conditions. Consider the control volumes for a lake and a river reach presented in Figure 6.2 (top and bottom, respectively). If we assume there's no inflow or outflow of groundwater in the pond, the bulk movement of mass will occur across the water surface. However, when we analyze the movement of constituents in a flowing river, mass can move in and out of the system through the interface with the atmosphere as well as from upstream to downstream reaches.

Once we've established a control volume, we need to track the rate at which mass moves into and out of the system. We also need to account for the creation or consumption of mass inside V_c (often through [bio]chemical reactions or sedimentation). If we add these components of the material balance, we can know the rate at which mass accumulates inside V_c.

From a practical point of view, it's helpful to have a rubric—a general set of steps to follow—when applying the material balance concept to real-life situations:

MATERIAL BALANCE RUBRIC

Step 1. Sketch the system. This doesn't have to be a work of art. Rather, it's a way to help visualize the physical system, its boundaries, and ultimately, the way materials move into and out of the system.

Step 2. List all pertinent data, including any underlying assumptions—known flowrates, concentrations, volumes, *etc.* Be sure to identify them on your sketch.

FIGURE 6.2 Control volumes for a lake (top) and a river reach (bottom).

Step 3. Identify the system boundary and show the control volume/reference frame on your sketch.

Step 4. Determine the basis for future calculations. This approach is commonly used to determine concentrations and flowrates. It can also be used to find the physical dimensions of a system or compartment.

Step 5. Write the material balance in words. This will make it easier to identify terms that can be eliminated from the mathematical expression. For example, if we need to find a steady-state concentration of a compound in water, we know that its concentration doesn't change over time.

Step 6. Write the mathematical expression for the material balance.

Step 7. Solve.

Example: The Movement of Mass into and out of a Lake

Let's apply the material balance rubric to examine the movement of water into and out of a lake to estimate the amount of water that accumulates over a year.

Step 1. Sketch the system (Figure 6.3). A simple sketch is a great way to visualize the movement of matter and energy in a system. Inflows to the lake include surface flows (runoff, rivers and streams, and even industrial discharges), groundwater, and precipitation. Outflows from the lake include surface flows, groundwater, and evaporation. In some cases, we might also include transpiration via plants.

Step 2. List all pertinent data (e.g., underlying assumptions, flow rates, concentrations, volumes, etc.) and identify them on your sketch as appropriate. The particular assumptions we use are driven by site-specific conditions. For example, if the pond was clay lined, we could neglect the loss or gain of water to/from the ground. We might also assume that no [bio]chemical reactions occur in the lake. This is generally a solid assumption unless you're told otherwise.

Step 3. Identify the system boundary and show the control volume on your sketch. When choosing a control volume, be sure to establish clear system boundaries and select a frame of reference that yields practical information. In this case, the lake itself would be an appropriate control volume. However, if we neglected

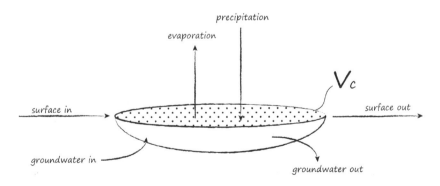

FIGURE 6.3 General inflows and outflows of water to/from a lake.

groundwater exchange, we might consider the control volume to be the water surface.

Step 4. Determine the basis for future calculations (time, the concentration of particular constituent, flow rate, *etc.*) In this example, we want to find a volume of water that accumulates in the lake over a year. As we'll see, the time frame we select can become an important factor in this analysis.

Step 5. Write the material balance in words. The general material balance statement was given in Figure 6.1. However, we can add more specificity to this expression by considering the exact ways water moves in the system (Equation 6.1). Water can enter the system as rain or snow. The pond can also receive water from overland flow/surface runoff. Water can be lost from this system through evaporation. If the pond contained aquatic vegetation, water could also be lost through evapotranspiration.

Equation 6.1 rate of water accumulated in V_c = rate of water into V_c – rate of water out of V_c +/– rate of water consumption or creation in V_c.

We can write Equation 6.1 in terms of the specific water sources and sinks (Equation 6.2). Since there are no other constituents in the pond, water doesn't undergo any (bio)chemical reactions, so there's no creation or consumption of water in the pond.

Equation 6.2 rate of water accumulated in V_c = (rate of precipitation into V_c + rate of surface water into V_c + rate of groundwater into V_c) – (rate of water evaporation from V_c + rate of surface water out of V_c + rate of groundwater out of V_c)

Step 6. Write the mathematical expression for the material balance. This step isn't nearly as straightforward as the others. For example, what's the best way to write the rate that water enters the lake as snow and rain? How do we account for the effect of seasonal temperature changes on the rate that water evaporates? Again, this is where the selection of an appropriate time interval can be critical.

Step 7. Solve. In practice, it can be difficult to quantify terms related to the rates of precipitation, runoff, evaporation, and evapotranspiration, as we saw in Chapter 3. To address this challenge, we often consider time periods on the order of years as opposed to months, weeks, or days when performing water balances. For example, on an annual basis, it might be appropriate to assume that precipitation and evaporation are roughly equal. Likewise, it could also be possible that the rate at which water that enters the lake from surface runoff is negligible when compared to other flows. In this case, if there's no groundwater exchange with the lake, there would be no water accumulation in the lake. However, immediately after a rainstorm, the rate that water enters the system is greater than the rate that water leaves the system, and water will accumulate in the lake. Since the boundary of the lake is fixed, the water level will rise. Consequently, our application of the material balance approach has to be informed by time in addition to the physical dimensions and concentrations of various constituents.

Let's continue this example by considering a 1×10^5 m³ lake with an average depth of 5 m that's fed and drained by a river (Figure 6.4). Assume the upstream river flows at a rate of 1×10^5 m³/year and contains a pollutant at a concentration of 10

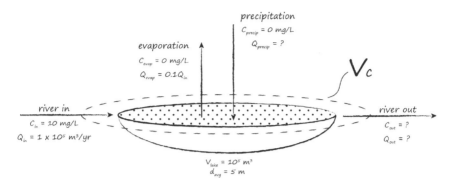

FIGURE 6.4   Schematic illustration of a lake that's fed and drained by a river.

mg/L. If 10% of the annual inflow from the river is lost to evaporation, what's the concentration of the pollutant that leaves the lake?

Step 1. Sketch the system (see Figure 6.4).

Step 2. List all pertinent data (see Figure 6.4). Known and unknown water flow rates and pollutant concentrations, as well as the lake volume and depth, are identified on the sketch.

One of the most difficult parts of these problems is getting started. For example, how should we account for the precipitation into the lake? Since we're not given data on precipitation, we need to find a source of this information that's appropriate for the specific case that's being analyzed. This is another substantial challenge for problem solvers. Sources of reviewed and defensible data on water quality, hydrology, weather, climate, and natural resources in the US are presented in Table 6.1. In this case, one approach is to use the fact that 767 mm of precipitation falls as rain and snow in the continental US each year (NOAA 2023). Just be sure to cite the source of your data.

Something else that we've tacitly assumed is that the pollutant doesn't evaporate. Consequently, the concentration of the pollutant in evaporation water is zero. Likewise, we have to assume that no additional pollutant enters the lake through precipitation and the pollutant doesn't react in the lake. These assumptions will help to simplify some of our calculations. At this point, we know or have assumed everything except the concentration of the pollutant in the lake discharge and the rate that water drains from the lake.

Step 3. Identify the system boundary and show the control volume on your sketch. The control volume includes the entire lake.

Step 4. Determine the basis for future calculations. We need to find the concentration of the pollutant that leaves the lake. To find this, we also need to determine the flow rate of water that leaves the lake, Q_{out}. Consequently, we'll need to perform a mass balance on water (a flow balance) and a mass balance on the pollutant. Note that the flows in this case are specified on an annual basis.

Step 5. Write the material balance expression(s) in words. We have to perform a flow balance and a mass balance on the pollutant.

A material balance on water:

The net flow of water into V_c is the sum of the upstream river flowrate and the rate that water enters the lake through precipitation. Similarly, the net rate that water leaves the lake is the sum of the downstream river flowrate and the rate that water evaporates from the lake.

A material balance on the pollutant:

The rate at which the pollutant enters the lake from upstream is equal to the rate the pollutant leaves the lake through the downstream river. The pollutant doesn't accumulate in the lake.

Step 6. Write the mathematical expression for the material balance of water and the pollutant (Equation 6.3 and Equation 6.4, respectively).

TABLE 6.1

Sources of Reviewed and Defensible Data on Water Quality, Hydrology, Weather, Climate, and Natural Resources in the US. Since the location of specific data sets can change over time, agency level websites are provided.

Agency	Website	Major Types of Data
US Department of Agriculture	www.usda.gov	Land use and land cover, high resolution aerial imagery, soil data (see NRCS)
Natural Resources Conservation Service	www.nrcs.usda.gov	Soil data, land use and cover data, aerial imagery
US Department of Commerce	www.commerce.gov	
National Institute of Standards and Technology	www.nist.gov	Data on the physical, chemical, and biochemical properties of materials
National Oceanic and Atmospheric Administration (NOAA)	www.noaa.gov	Geospatial data, weather and climate data (see NOAA)
National Weather Service		
National Weather Service	www.nws.gov	National and local weather data, climate data, hydrologic data
US Department of the Interior	www.doi.gov	
US Bureau of Reclamation	www.usbr.gov	Hydrologic data for the western US
US Geological Survey	www.usgs.gov	Geological data, hydrologic data, data on biota
US Fish and Wildlife Service	www.fws.gov	Data on biota including the National Wetlands Inventory
National Aeronautics and Space Administration	www.nasa.gov	Remote sensing data, climate data
US Environmental Protection Agency	www.epa.gov	Analytic procedures, properties of air and water pollutants, solid waste management, hazardous waste management, inventories of contaminated sites
State geologic, water, and natural history surveys		State and local-level data on soils, geology, water, and biota. These data are often shared with federal agencies and are incorporated onto those databases

$$\text{Equation 6.3} \quad Q_{in} + Q_{precip} - Q_{out} - Q_{evap} = 0$$

$$\text{Equation 6.4} \quad C_{in}Q_{in} + C_{precip}Q_{precip} - C_{out}Q_{out} - C_{evap}Q_{evap} = 0$$

Step 7. Solve.

Before we can calculate Q_{out}, we need to find Q_{precip} and Q_{evap}. Since we know the volume of the lake and the average depth, we can find the cross-sectional area (Equation 6.5). Then, we can determine the annual precipitation (Equation 6.6) as the product of the area and the annual precipitation rate.

$$\text{Equation 6.5} \quad A = \frac{V}{d} = \frac{10^5 m^3}{5m} = 2 \cdot 10^4 m^2$$

$$\text{Equation 6.6} \quad Q_{precip} = \left(0.767\frac{m}{yr}\right)A = \left(0.767\frac{m}{yr}\right)\left(2 \cdot 10^4 m^2\right) = 1.53 \cdot 10^4 \frac{m^3}{yr}$$

We were told that Q_{evap} is one tenth of Q_{in} (Equation 6.7).

$$\text{Equation 6.7} \quad Q_{evap} = 0.1 Q_{in} = (0.1)\left(10^5 \frac{m^3}{yr}\right) = 10^4 \frac{m^3}{yr}$$

Using Equation 6.7, $Q_{out} = 1.05 \cdot 10^5$ m³/yr.

Now, we can find the pollutant concentration in the downstream river, C_{out}, from Equation 6.4. Before solving for C_{out}, we should have another look at Equation 6.4 in relationship to our assumptions to see if we can simplify any terms. In this case, $C_{precip} = C_{evap} = 0$ and the mass balance on the pollutant can be written in a much simpler form (Equation 6.8).

$$\text{Equation 6.8} \quad C_{in}Q_{in} - C_{out}Q_{out} = 0$$

Now, we can find that $C_{out} = 9.52$ mg/L (Equation 6.9).

$$\text{Equation 6.9} \quad C_{out} = \frac{C_{in}Q_{in}}{Q_{out}} = \frac{\left(10\frac{mg}{L}\right)\left(10^5 \frac{m^3}{yr}\right)}{1.53 \cdot 10^4 \frac{m^3}{yr}} = 9.52\frac{mg}{L}$$

Example: Adding More Real-World Significance to the Mass Balance on the Lake

In the preceding example, we neglected water movement through transpiration. Would the concentration in the downstream river be higher, lower, or unchanged if we included the transpiration flow of water, Q_t?

The net flow of water in the downstream river when transpiration is included, Q_{out+t} (Equation 6.10), is lower than Q_{out}.

$$\text{Equation 6.10} \quad Q_{out-t} = Q_{in} + Q_{precip} - Q_{evap} - Q_t$$

As in the case of precipitation and evaporation, $C_t = 0$ mg/L, and the expression for C_{out+t} is given in Equation 6.11 since $Q_{out+t} < Q_{out}$ and $C_{out+t} > C_{out}$.

$$\text{Equation 6.11} \quad C_{out+t} = \frac{C_{in}Q_{in}}{Q_{out-t}}$$

Example: The Origins of the Ohio River—Material Balances to Find Flow and Sediment Concentrations

The Allegheny and Monongahela Rivers meet in Pittsburgh to form the Ohio River (Anderson 2000). The Allegheny River flows from the north through forests and small towns with an average flow of 580 m^3/min and a sediment load of 250 mg/L. The Monongahela River flows from the south at 780 m^3/min through old steel towns, active and closed surface and underground mines, and some farmland. The Monongahela River carries a sediment concentration of 1,500 mg/L. Let's use the material balance approach to find the average sediment concentration immediately below the confluence of the Ohio River.

Steps 1, 2, and 3. Sketch the system (see Figure 6.5).

Step 4. Establish a basis for future calculations. We know the flow rates and silt concentrations of both "feeder" rivers, but we don't know either parameter for the Ohio River. To find the silt concentration in the Ohio River, we need to determine the flow rate of water after the confluence of the two rivers. We can perform a flow balance at the confluence of the three rivers to find the flow rate in the Ohio

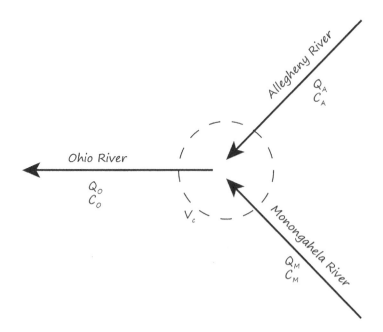

FIGURE 6.5 Sketch of the confluence of the Allegheny, Monongahela, and Ohio Rivers in Pittsburgh, PA.

River. Then, we can find the sediment concentration in the Ohio by performing a mass balance on sediment at the confluence.

Step 5. Write the general word-based mass balance statement. In this system, sediment doesn't react and there's no accumulation of water or sediment in the control volume. Consequently, the rate that sediment flows into the control volume from the Allegheny and Monongahela Rivers must be equal to the rate that sediment flows out of the control volume in the Ohio River.

Step 6. Write the mathematical mass balance statement.

We need to begin with a flow balance to find Q_O. Since there are no reactions or accumulation of sediment in the control volume, the flow balance is given in Equation 6.12.

$$\text{Equation 6.12} \quad Q_O = Q_A + Q_M$$

Now we can write a mass balance equation that relates the sediment concentrations and water flow rates of each river (Equation 6.13).

$$\text{Equation 6.13} \quad C_O Q_O = C_A Q_A + C_M Q_M$$

$$\text{Equation 6.14} \quad C_O = \frac{C_A Q_A + C_M Q_M}{Q_O}$$

Step 7. Solve. The average flow rate of the Ohio River is 1,360 m³/min (from Equation 6.12) and the average silt concentration in the Ohio River is 967 mg/L (Equation 6.14).

6.2   MECHANICAL ENERGY IN WATER

As noted previously, the material balance approach can also be used to account for the presence and transformation of energy in a discrete system. While energy can take many forms (thermal, electrical, chemical, *etc.*), we're going to be most concerned with mechanical energy in flowing water systems. In flowing water systems, there is a dynamic exchange between potential and kinetic energy. Potential energy is frequently called energy of position—the energy a body has by virtue of its location relative to some place with lower energy. We can write the potential energy, E_p, as the product of the gravitational force, F_g, acting on an elevated mass, m, at some height, Δh, as shown in Equation 6.15

$$\text{Equation 6.15} \quad E_p = F_g \Delta h = mg\Delta h$$

where the acceleration due to gravity, g, is 9.81 m/s² or 32.2 ft/s².

In some cases, a common point such as mean sea level is used as a reference, while in other cases, a local datum such as the elevation of a water control structure is used.

In a number of disciplines, water pressure is often referred to in terms of the force exerted by a column of water, which is known as the "pressure head." In this case, water pressure is represented in units of "ft of water" (ft H_2O). Due to its ubiquitous use, this is one case where SI units will be eschewed in favor of standard units. Similar to gravitational potential energy, pressure head is related to the position or elevation of the water column above some point. Consider a cube of water that's 1 inch wide, deep, and long. The area of the base is 1 in^2 and the water depth is 1 inch. When gravity acts on the mass of water in the 1 in^3 of water in the container, a force of 0.0361 lb$_f$ is exerted on the bottom face of the cube. In this case, lb$_f$, refers to pounds of force and not pounds of mass. Since the base area of the container is 1 in^2, the pressure exerted by a 1 in high water column is 0.0361 lb$_f$/in^2 (psi). When 12 containers are stacked on one another, the pressure exerted at the bottom of the 1 ft tall water column is 0.433 psi. Alternatively, we can say that 2.31 ft of water exerts a pressure of 1 psi (~6.90 kPa).

In contrast to potential energy, kinetic energy is energy of motion. We can determine the kinetic energy, E_k, of a mass with a velocity, v, according to Equation 6.16.

$$\text{Equation 6.16} \quad E_k = \frac{1}{2}mv^2$$

Under ideal conditions, all environmental factors remain constant, objects have homogeneous properties, and we can neglect the impacts of friction. While we know that these constraints aren't reflected in real life, they provide a set of simplified circumstances that can serve as a place to begin understanding how these parameters impact real situations. For example, consider a lake that drains into a stream located some elevation below the discharge point. Assume there's no friction and that water proceeds from the lake to the stream without any turbulence. By virtue of its elevation, water in the lake possesses gravitational potential energy that's proportional to its mass and its distance above the stream bed. When water enters the stream from above, its potential energy has been converted into kinetic energy and the water continues to flow to points of lower elevation.

In reality, the transformation of potential energy into kinetic energy in this system is more complicated. For example, as water falls from the discharge to the stream, it experiences a drag force due to the resistance of air. Likewise, when water reaches the streambed, energy is dissipated on rocks and in the channel. Regardless, no energy has been created or destroyed in either system. These particular energy sinks are addressed in more detail when we consider the movement of particles in a viscous fluid.

Example: Assessing the Impacts of Drought on Water for Hydropower Production

Lake Mead is the largest manmade water reservoir in the United States. Under non-drought conditions, the surface area of Lake Mead is around 590 km^2 with a depth of 160 m. Since the early 2000s, the Colorado River Basin has experienced an unprecedented period of drought that has reduced the water level in the lake to around 25% of its "full" depth. Along with a reduction in water depth, assume the surface area of the lake has been reduced by 40% under drought conditions (Forsythe et al. 2012) (Figure 6.6). Determine the difference in potential energy when Lake Mead is full and under drought conditions.

In this case, the change in potential energy of water in Lake Mead is due to the lower elevation of the waterbody and the decrease in the mass of water

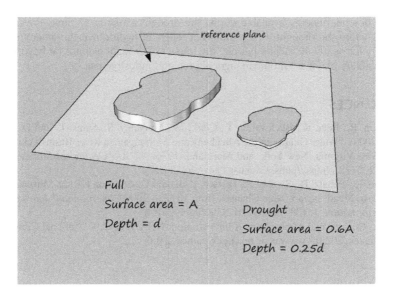

FIGURE 6.6 Lake Mead depth and surface areas at full and drought stages.

that's held in the reservoir. To determine the potential energy, we need to know the mass of the water in Lake Mead, which we can find as the product of the lake's volume and the density of water. However, we first need to find the mass of water in the lake using the relationship between density, mass, and volume. We can use the information we're given as a starting point to find the physical dimensions of the lake. Based on the information we've been given, the mass of water available in Lake Mead under non-drought conditions can be found using Equation 6.17.

$$\text{Equation 6.17} \quad m_{water} = V_{lake}\rho_{water} = A_{lake}d_{lake}\rho_{water}$$

$$= \left(590 \cdot 10^6\, m^2\right)\left(160m\right)\left(10^3\, \frac{kg}{m^3}\right) = 9.44 \cdot 10^{13}\, kg$$

Be careful when converting km² to m² in Equation 6.17. One km² is equal to 10^6 m²—not 10^3 m². At a depth of 160 m, potential energy of the water in the Lake is $1.48 \cdot 10^{17}$ J (Equation 6.18).

$$\text{Equation 6.18} \quad \cdot E_p = \left(9.44 \cdot 10^{13}\, kg\right)\left(160m\right)\left(9.81\frac{m}{s^2}\right) = 1.48 \cdot 10^{17}\, J$$

Under drought conditions, the surface area is 40% smaller, the depth is 75% less. This corresponds to a water volume of $1.42 \cdot 10^{10}$ m³ and a water mass of $1.42 \cdot 10^{13}$ kg. At an elevation of 40 m, the potential energy available is $5.56 \cdot 10^{15}$ J (Equation 6.19).

$$\text{Equation 6.19} \quad E_p = \left(1.42 \cdot 10^{13}\, m^3\right)\left(0.25\right)\left(160m\right)\left(9.81\frac{m}{s^2}\right) = 5.65 \cdot 10^{15}\, J$$

The potential energy available during the drought is nearly 27 times lower than that before the drought. In a region that relies on hydropower to generate electricity for nearly 40 million people, the impact of drought extends far beyond the availability of water for human, agricultural, and industrial use.

REFERENCES

Anderson, R., Beer, K., Buckwalter, T., Clark, M., McAuley, S., Sams, J., and D. Williams (2000). "Water Quality in the Allegheny and Monongahela River Basins Pennsylvania, West Virginia, New York, and Maryland, 1996–98," *US Geological Survey Circular*, 1202 (32), https://pubs.water.usgs.gov/circ1202/.

Forsythe, K., Schatz, B., Swales, S., Ferrato, L.-J., and D. Atkinson (2012). "Visualization of Lake Mead Surface Area Changes from 1972 to 2009," *International Journal of Geo-Information*, 1, 108–119, doi:10.3390/ijgi1020108.

National Oceanic and Atmospheric Administration (NOAA) (2023). "National Climatic Data Center (2023)," www.ncdc.noaa.gov/oa/ncdc.html.

7 Forces Acting in Fluids

To describe the flow of fluids such as water, we need to relate material properties of the fluid to operating conditions and the geometry of the flow channel. Let's consider a cube with characteristic length, L, that's filled with a fluid that has a density, ρ, and viscosity, μ.

From Newton's second law of motion, we can write the force acting on a body, F, as the product of its mass, m, and acceleration, a (Equation 7.1). Recall that vector quantities are fully described by a magnitude and direction.

$$\text{Equation 7.1} \quad \vec{F} = m\vec{a}$$

We can use this basic relationship to consider individual forces acting on the fluid element. In particular, we'll need to know the gravitational, inertial, and viscous forces acting on/in our element.

7.1 GRAVITATIONAL FORCE

The force of gravity, F_g, acting on a body of mass, m, is given in Equation 7.2.

$$\text{Equation 7.2} \quad \vec{F_g} = m\vec{g}$$

Based on the information we know about the cube and its contents, we can find the mass of the fluid, m_{fluid}, from the density, ρ_{fluid}, and volume, V_{fluid}, which in this case is L^3 (Equation 7.3).

$$\text{Equation 7.3} \quad m_{fluid} = \rho_{fluid} V_{fluid} = \rho_{fluid} L^3$$

Now, we can write the force of gravity according to Equation 7.4.

$$\text{Equation 7.4} \quad \vec{F_g} = \rho_{fluid} L^3 \vec{g}$$

Since the force of gravity is directed downward, the vector notation is often dropped for simplicity.

7.2 INERTIAL FORCE

By virtue of having mass, objects have an inherent resistance to any change in their state of motion. Objects at rest tend to remain at rest, while objects in motion remain in motion unless they're acted on by an external force. The instantaneous acceleration, a, of a body moving with a velocity, v, at time, t, can be determined from Equation 7.5.

$$\text{Equation 7.5} \quad \vec{a} = \frac{\vec{v}}{t}$$

DOI: 10.1201/9781003289630-8

Likewise, the instantaneous velocity can be found from a body's position, x, over time as in Equation 7.6.

$$\text{Equation 7.6} \quad \vec{v} = \frac{\vec{x}}{t}$$

Using Equation 7.5 and Equation 7.6, we can express acceleration in terms of velocity and position as in Equation 7.7.

$$\text{Equation 7.7} \quad \vec{a} = \frac{\vec{v}^2}{\vec{x}}$$

In this case, the position, x, corresponds to the cube length, L. This allows us to write the inertial force, F_i, according to Equation 7.8 (Bird et al. 1960).

$$\text{Equation 7.8} \quad \vec{F_i} = m\frac{\vec{v}^2}{\vec{x}} = \rho_{fluid} L^2 \vec{v}^2$$

7.3 VISCOUS FORCE

Viscosity is a measure of a fluid's resistance to the shear stress induced by flow. In fluid mechanics, there are two types of viscosity: (1) absolute viscosity, μ, which is a measure of a fluid's internal resistance to flow and (2) kinematic viscosity, ν, which is the ratio of the absolute viscosity to the fluid density. This reflects the momentum of molecules in the fluid. In Newtonian fluids like water, the stresses that result from the fluid's flow are linearly related to the shear rate when the temperature and pressure are constant. In general, Newtonian fluids include low molecular weight compounds like water, light crude oil, and gases, as well as solutions of low molecular weight compounds. Changes in the viscosity of Newtonian fluids can be made by altering the fluid's temperature or pressure.

A viscous force, F_v, (Equation 7.9) results when a shear stress, τ, (Equation 7.10) is exerted between fluid layers. For water flowing in a stream, the viscous force is proportional to the fluid velocity, the area over which shear stress is exerted (A), and the fluid viscosity (μ). Conversely F_v is inversely proportional to the fluid depth (see Equation 7.11) (Bird et al. 1960).

$$\text{Equation 7.9} \quad \vec{F_v} = \vec{\tau} A$$

$$\text{Equation 7.10} \quad \vec{\tau} = \frac{\mu\vec{v}}{L}$$

$$\text{Equation 7.11} \quad \vec{F_v} = \frac{\mu\vec{v}}{L} A = \mu\vec{v}L$$

7.4 HYDRAULIC TURBULENCE

Before we discuss specific mass and energy transport mechanisms, we need to consider the role of hydraulic turbulence and the manner in which it influences the downstream movement of water and any entrained constituents. Turbulence comes from

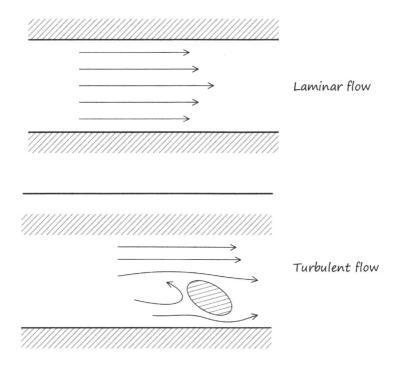

FIGURE 7.1   Stream flow lines for laminar (top) and turbulent (bottom) flow.

irregular, uneven motion in a fluid (a liquid or a gas). When turbulence is low, water molecules move downstream in parallel paths at a constant velocity. This is known as laminar flow (see Figure 7.1, top). In contrast, under turbulent conditions, water molecules follow different paths that have different velocities (see Figure 7.1, bottom). Turbulent flow occurs at elevated water velocities and when water encounters obstructions to downstream flow. In natural streams, typical obstructions include boulders, woody debris, and irregularities in the stream bed or banks. Under turbulent flow, mixing is considered high.

The Reynolds number (Re) is a dimensionless parameter that we use to characterize the hydraulic turbulence in a flowing fluid—this includes both liquids and gases. Regardless of the particular fluid or geometry being considered, the Reynolds number is defined as the ratio of inertial to viscous forces or resistances (Equation 7.12).

$$\text{Equation 7.12} \quad Re = \frac{F_i}{F_v}$$

Factors that resist changes in the speed and/or direction of water flow are known as inertial resistances. These resistances arise in response to the force of the moving fluid. You will recall that inertia is the tendency of an object to resist any change in its velocity unless acted upon by an external force. Since acceleration is a change in velocity over time, the inertial force, F_i, can be determined from an object's mass and velocity. Explicit forms of the inertial force in discrete and differential forms are presented in Equation 7.13 and Equation 7.14. If a body is moving at a constant velocity (including when it's at rest), F_i is zero because the velocity doesn't change over time. To keep things simple, the vector notation has been dropped and we'll consider motion in only one direction, such as the flow of a fluid in a channel or pipe.

$$\text{Equation 7.13} \quad F_i = m\left(\frac{\Delta v}{\Delta t}\right) = ma$$

$$\text{Equation 7.14} \quad F_i = m\left(\frac{dv}{dt}\right) = ma$$

The instantaneous acceleration, a, velocity, v, and position, x, are related to time in Equation 7.15.

$$\text{Equation 7.15} \quad a = \frac{v}{t} = \frac{v^2}{x}$$

This allows us to write a more useful form of the inertial force (Equation 7.16):

$$\text{Equation 7.16} \quad F_i = \frac{mv^2}{x}$$

In contrast to inertial forces, viscous forces are created from friction between water molecules. The viscous force is the product of the shear stress, τ (Equation 7.10), that develops between fluid layers and the area over which the shear stress is applied.

The absolute viscosity and density of water at a range of environmentally relevant temperatures are presented in Table 7.1 (Haynes 2014).

This gives us a form of the Reynolds number that can be calculated from data on the properties of the fluid and geometry of the channel (see Equation 7.17).

$$\text{Equation 7.17} \quad Re = \frac{\rho x^2 v^2}{\mu v x} = \frac{\rho x v}{\mu}$$

TABLE 7.1
Absolute Viscosity and Density of Water at a Variety of Environmentally Relevant Temperatures (Haynes 2014).

Temperature		Absolute Viscosity , cP	Density*, kg/m³
°F	°C		
32	0	1.794	999.87
40	4.4	1.546	999.99
50	10.0	1.310	999.75
60	15.6	1.129	999.07
70	21.1	0.982	998.02
80	26.7	0.862	996.69
90	32.2	0.764	995.10
100	37.8	0.682	993.18

* At standard atmospheric pressure.

When using dimensionless numbers, the scale of a system is defined by a character-istic length. In fluid mechanics, the characteristic length is the hydraulic radius, R_h, (Equation 7.18) which incorporates both the cross-sectional area of the fluid and the wetted perimeter, p_w, of the flow channel, duct, or pipe.

$$\text{Equation 7.18} \quad R_h = \frac{A}{p_w}$$

The wetted perimeter is the length of the water cross section that's in contact with the channel or pipe. Using the characteristic length is especially useful when using dimen-sionless numbers, as we know that the cross-sectional geometry of channels, ducts, and pipes can vary. The cross-sectional areas, hydraulic radii, and wetted perimeters for three different channel geometries are shown in Figure 7.2. If we replace x, the length variable in the Reynolds number equation, with the hydraulic radius, we can write Re in a way that incorporates channel dimensions (Equation 7.19).

$$\text{Equation 7.19} \quad Re = \frac{\rho R_h v}{\mu}$$

Also, since we're using the Reynolds number to represent the turbulence in bulk flow-ing water, we use the average velocity when making calculations with Equation 7.19.

The Reynolds number at which flow transitions from laminar to turbulent condi-tions is a function of the system being examined. In open channels (waterbodies that are open to the atmosphere and flow under the influence of gravity), laminar flow conditions prevail when Re < 500. In this case, Re is determined primarily by water viscosity. Since the viscosity of water is low relative to that of other fluids, small irregu-larities in the stream bed are transmitted with relative ease. In contrast, turbulent flow occurs when Re > 2,000. Under turbulent flow conditions, Re is controlled by water velocity. Reynolds numbers between 500 and 2,000 for open channel flow are indica-tive of a transition between laminar and turbulent conditions (Sturm 2010). In this case, flow is a combination of laminar and turbulent conditions. As you'll begin to notice when working through problems, turbulent conditions are common in flowing water systems. In contrast to open channel flow, laminar flow in a pressurized pipe occurs when Re < 2,000 and turbulent conditions occur at Re > 4,000 (Bird *et al.* 1960).

7.5   FROUDE NUMBER

The Froude number, Fr, is a dimensionless parameter that relates flow velocity to the flow depth and gives us an idea of how energy in a flowing waterbody is transmitted in relationship to channel dimensions. Specifically, the Froude number is the ratio of inertial forces to gravitational forces acting on a system (Equation 7.20). In this con-text, the denominator represents the speed of a small wave that travels on the water surface and the numerator is the average speed of the bulk flowing water.

$$\text{Equation 7.20} \quad Fr = \frac{v}{\sqrt{gy}}$$

where v = average water speed, g = gravitational constant, and y = average water depth.

A rectangular channel.

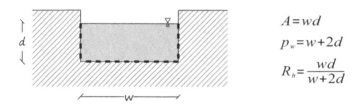

$$A = wd$$

$$p_w = w + 2d$$

$$R_h = \frac{wd}{w + 2d}$$

A trapezoidal channel.

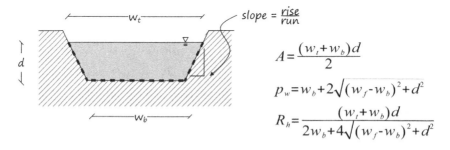

$$slope = \frac{rise}{run}$$

$$A = \frac{(w_t + w_b)d}{2}$$

$$p_w = w_b + 2\sqrt{(w_f - w_b)^2 + d^2}$$

$$R_h = \frac{(w_t + w_b)d}{2w_b + 4\sqrt{(w_f - w_b)^2 + d^2}}$$

A triangular channel.

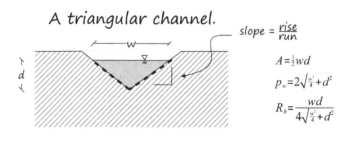

$$slope = \frac{rise}{run}$$

$$A = \tfrac{1}{2}wd$$

$$p_w = 2\sqrt{\tfrac{w^2}{4} + d^2}$$

$$R_h = \frac{wd}{4\sqrt{\tfrac{w^2}{4} + d^2}}$$

- - - - - - - - wetted perimeter

FIGURE 7.2 The cross-sectional areas, hydraulic radii, and wetted perimeters for rectangular, trapezoidal, and triangular channels.

If you drop a pebble in a quiescent pool, concentric waves will radiate from the point at which the pebble enters the water with a velocity, v_{wave}, as in Case A (Figure 7.3). Now, we need to consider what happens when the pebble is dropped into flowing water and $Fr = 1$ (Case B, Figure 7.3). This condition is known as critical flow and occurs when the inertial and gravitational forces are balanced. The velocity and water depth at critical flow are known as the critical depth and critical velocity.

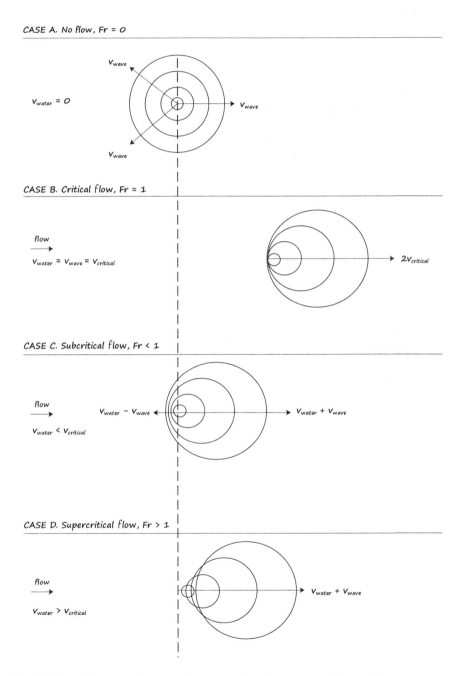

FIGURE 7.3   Flow interference patterns at a variety of water velocities and Froude numbers.

At critical flow conditions, the speed of the small surface wave, and the speed of bulk water are equal and the wave forms that were caused by the pebble move downstream with a velocity equal to $2v_{critical}$ as shown in Figure 7.3.

When the velocity of the water is less than $v_{critical}$, the wave form created by the disturbance possesses enough inertia for a fraction of the wave to be to be transmitted above the point of the disturbance, as in Case C (Figure 7.3). This is known as subcritical flow, and part of the waveform is directed back upstream. When the average bulk water velocity is high and the water is relatively shallow, Fr > 1, and flow is supercritical. Under super critical conditions, the velocity of the bulk fluid is greater than the velocity of the surface wave and disturbances are transmitted downstream in the direction of bulk water flow, as shown in Case D (Figure 7.3). Under supercritical flow conditions, flow disturbances are not transmitted back upstream (Sturm 2010).

7.6 HYDRAULIC JUMP

The transition from supercritical to subcritical flow conditions is known as a hydraulic jump. As a result of the rapid increase in water depth and/or decrease in water velocity, a standing wave is generated. When the standing wave encounters slower moving water, its momentum slows and the height of the water column increases (see Figure 7.4). This phenomenon results in a significant increase in air entrainment and turbulence in the vicinity of the hydraulic jump (Rouse *et al.* 1959).

Hydraulic jumps are commonly found in the vicinity of boulders or woody debris, and in places where there are abrupt changes in channel geometry. Weirs, spillways, and waterfalls are also common sites where hydraulic jumps occur. In traditional river and stream restoration, there has been a focus on channel shape and substrate characteristics. However, there are significant opportunities to integrate

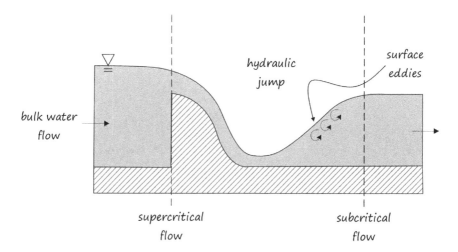

FIGURE 7.4 Schematic of a hydraulic jump at the transition between supercritical and subcritical flow.

these findings with new knowledge on the energy dissipating abilities of natural, in-stream structures created by large deposits of woody debris and boulders.

7.7   REYNOLDS NUMBER AND FROUDE NUMBER

There are a number of similarities and differences between the Reynolds number and Froude number. Both expressions contain the average water velocity and provide some measure of the amount of energy that's available to entrain and subsequently transport sediment. Under conditions of high hydraulic turbulence, sediment can be maintained in the water column. When hydraulic turbulence is low, sediment is more likely to settle from the water column. Unlike the Reynolds number, the Froude number relates the average water velocity to the average depth of the water column and can be used to predict the effects of sudden changes in channel geometry. Expressions for Re and Fr also contain length terms, R_h and y, respectively. While the hydraulic radius and average water depth are not equivalent and cannot be used interchangeably, there are circumstances when they approximate one another. For example, the hydraulic radius and average depth are approximately equal in a wide and shallow rectangular channel.

7.8   THE MOVEMENT OF MASS

7.8.1   ADVECTION

Advection is the transport of a substance along with the bulk flow of a fluid. The downstream movement of suspended sediment in a stream is an example of advective transport. In contrast, convection generally refers to the movement of a fluid under the influence of a temperature gradient. Temperature stratification in lakes results in differences in water density between the epilimnion and hypolimnion which can induce a convective current. In natural systems such as lakes, advection typically occurs in the horizontal direction, while convection occurs in the vertical direction.

Consider a rectangular channel that contains flowing water moving with a veloc-ity, v. Equations for water velocity in discrete and differential forms are presented in Equation 7.21 and Equation 7.22, respectively.

$$\text{Equation 7.21} \quad \vec{v} = \frac{\Delta \vec{x}}{\Delta t}$$

$$\text{Equation 7.22} \quad \vec{v} = \frac{d\vec{x}}{dt}$$

where x is position and t is time.

Since velocity is a vector, it's direction must also be specified. However, in simple cases, the direction of the velocity vector is parallel to the direction of water flow and vector notation is often dropped.

The velocity of water is affected by its position in the water column, channel bed and bank roughness, channel geometry, and channel slope. For example, consider the water velocity profiles for a uniform rectangular channel presented in Figure 7.5. In the plan view (Figure 7.5, top), the highest water velocity occurs at the middle of the

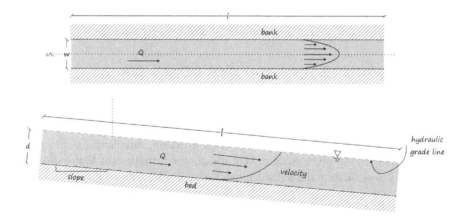

FIGURE 7.5   Water velocity profiles in plan (top) and profile (bottom) views.

stream and lowest water velocities occur at each of the two stream banks. Similarly, when viewed in profile (Figure 7.5, bottom), water velocity is lowest at the stream bed and highest at the water surface. The asymmetric velocity distribution arises from friction between water and the stream bed and banks. It's important to note that at the stream banks, the water velocity and the velocity of the stream bed/bank are both zero. This is known as the zero-slip condition (Sturm 2010).

7.8.2   Channel Dimensions

Since we know that the cross-sectional geometry of stream channels can vary, the hydraulic radius, R_h, is used to describe the shape of a stream channel (Equation 7.18).

In the expression for the hydraulic radius, the wetted perimeter, p_w, is the length of the water cross section that comes in contact with the channel bed and the banks. The free water surface is not included when determining the wetted perimeter. In the case of a uniform rectangular channel, p_w is given by Equation 7.23.

$$\text{Equation 7.23} \quad p_w = d + w + d = 2d + w$$

The hydraulic radius can also be used as a basis to compare the overall efficiency of water and sediment transport between two or more streams and/or the same stream at different stages (water depths). R_h depends on the geometry of the channel so the explicit forms for cross-sectional area and wetted perimeter must be determined on a case-by-case basis. For the rectangular stream section, the corresponding hydraulic radius is given in Equation 7.24.

$$\text{Equation 7.24} \quad R_h = \frac{wd}{2d + w}$$

Since friction between flowing water and the stream bed and banks results in a loss of energy that's manifested by a decrease in water velocity, the wetted perimeter and by extension, the hydraulic radius, can provide valuable information on a channel's overall efficiency.

Example: The Impact of Channel Dimensions on Key
Hydraulic Parameters – Channels with Equal Area

Consider two channels with uniform rectangular cross sections of equal area but different widths and depths (Table 7.2).

In this case, the cross-sectional area is 12 m² for both channels; however, the hydraulic radius of the narrow and deep channel has a 40% greater R_h for the wide and shallow channel. Consequently, the narrow and deep channel is more efficient at moving water and sediment.

Now, let's consider how the water depth impacts the hydraulic radius and consequently, the efficient movement of water and sediment. In this case, we can consider a stream that has a seasonal low water flow and when the channel is running full. For example, consider the 4 m wide channel from the previous discussion at water depths of 0.2 m and 3 m. Based on the calculations in Table 7.3, deep water is more than ~5.7 times more efficient at moving water and sediment.

TABLE 7.2
Hydraulic Radius and Wetted Perimeter for Two Uniform Rectangular Channels with Equal Cross-Sectional Areas but Different Width and Depth.

Width, m	Depth, m	Channel Description	A, m²	p_w, m	R_h, m
12	1	Wide and shallow	12	14	0.86
4	3	Narrow and deep	12	10	1.20

TABLE 7.3
Hydraulic Radius and Wetted Perimeter for A 4 M Wide Channel at Two Water Depths.

Width, m	Depth, m	Channel Description	A, m²	p_w, m	R_h, m
4	3	Deep water	12	10	1.20
4	0.2	Shallow water	0.8	4.4	0.18

7.8.3 Water Velocity

Often, the average water velocity is considered representative of stream flow and is used in many calculations. As we saw previously, the velocity of water in open channel flow is a function of the channel geometry. We also know that the material properties of the channel bed can have an impact on water velocity. For example, when compared to a channel with a smooth surface, a channel that has a rough, irregular surface dissipates more energy. Manning's equation (Equation 7.25) is among the most common tools used to determine the average water velocity in an open channel.

$$\text{Equation 7.25} \quad v = \left(\frac{\alpha}{n}\right)\left(R_h\right)^{2/3}\left(s\right)^{1/2}$$

where $\alpha = 1.0$ for SI units and $\alpha = 1.486$ for standard units; n = Manning's roughness coefficient (see Table 7.4), and s = slope. Since Manning's equation is an empirical relationship, it's important to be certain of the units of measurement that are used. When SI units are used, the velocity is given in m/s, R_h is in meters, and slope is in m/m. Likewise, in standard units, the velocity is given in ft/s, R_h is in ft, and slope is in ft/ft.

In Equation 7.25, Manning's roughness coefficient is a function of the stream bed material. Manning's roughness coefficients for several common stream bed materials are presented in Table 7.4 (Chow 1959; Arcement 1989; Sturm 2010). By introducing the channel's cross-sectional area, A, Manning's equation can be extended to find the average discharge from a channel, Q, as in Equation 7.26.

$$\text{Equation 7.26} \quad Q = vA = A\left(\frac{\alpha}{n}\right)\left(R_h\right)^{2/3}\left(s\right)^{1/2}$$

Manning's equation is valid for turbulent flow. While turbulent flow occurs readily in flowing water systems under typical environmental conditions, to be rigorous, the Reynolds number should be determined to verify the flow regime before applying Manning's equation.

TABLE 7.4
Manning's Roughness Coefficient, n, for a Variety of Channel Bed Materials and Conditions (Arcement 1989) for use in Equation 7.25 and Equation 7.26.

Material/Channel Conditions	n
Main channel of a natural stream	
clean, straight, full stage, no deep pools	0.025–0.033
clean, winding, some pools	0.033–0.045
sluggish reaches, weedy, deep pools	0.050–0.080
Concrete-lined channel	
float finished	0.011–0.015
unfinished	0.014–0.020
gravel bottom with riprap sides	0.023–0.036
Excavated or dredged channel	
earth, straight, and uniform	0.016–0.020
clean gravel, uniform section	0.022–0.030
short grass, few weeds	0.022–0.033

Example: Using Manning's Equation to Find
Channel Dimensions (in Standard Units)

Find the depth of water in a clean, straight, stream with no deep pools that can be approximated as a rectangular channel with a bottom slope of 0.0003 and a flow rate of 15 ft³/s. The channel width is about three times its depth. To find the depth, we need to relate the flowrate to channel dimensions. We know that the flow rate is the product of the cross-sectional area and the water velocity (Equation 7.26). Since we're told that the stream width is three times the depth, we can express the flowrate in terms of depth.

$$\text{Equation 7.27} \quad Q = vA = vwd = (3d^2)v$$

In this case, we can use Manning's Equation (in standard units) for open channel flow, assuming flow is uniform and steady. Since the channel is "a clean, straight, stream with no deep pools," Manning's coefficient, n, ranges from 0.025 to 0.033 (Table 7.4). For further calculations, we'll use the average n value of 0.029 and we can include Manning's Equation variables in our equation for flowrate (Equation 7.28).

$$\text{Equation 7.28} \quad Q = vA = vwd = (3d^2)\left(\frac{1.49}{0.029}\right)(R_h)^{2/3}(0.0003)^{1/2}$$

Before we can proceed, we need to recognize that the hydraulic radius also depends on channel dimensions (Equation 7.29).

$$\text{Equation 7.29} \quad R_h = \frac{3d^2}{2d+w} = \frac{3d}{5}$$

$$\text{Equation 7.30} \quad Q = (3d^2)\left(\frac{1.49}{0.029}\right)\left(\frac{3d}{5}\right)^{2/3}(0.0003)^{1/2} = 15\frac{ft^3}{s}$$

After gathering all of the constants and given data on one side and the terms that contain the depth on the other, we're left with the expression in Equation 7.31. This can be solved using a graphing calculator or a spreadsheet.

$$\text{Equation 7.31} \quad (3d^2)\left(\frac{3d}{5}\right)^{2/3} = 15\left(\frac{0.029}{1.49}\right)\left(\frac{1}{(0.0003)^{1/2}}\right) = 16.86$$

The depth of the channel is 2.17 ft, which in the context of this problem is ~2.2 ft.

7.8.4   SEDIMENTATION

Sedimentation is the movement of particles in water under the force of gravity. To develop an understanding of the physics involved in sedimentation, it is useful to examine the simplest case of a single particle that settles without any interaction with other particles (see Figure 7.6). For a particle to settle under gravity, there must be a

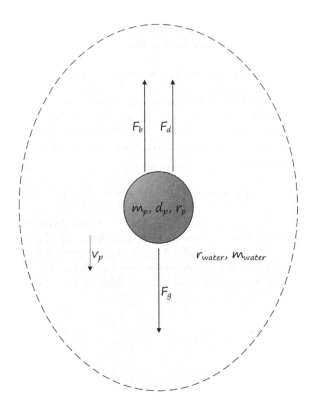

FIGURE 7.6   Force balance on a single particle settling in water.

net force, F_{net}, acting on it. To find the net force acting on the particle, we can perform a force balance as in Equation 7.32.

$$\text{Equation 7.32}\quad \overrightarrow{F_g} - \overrightarrow{F_b} - \overrightarrow{F_d} = \overrightarrow{F_{net}}$$

where F_g is the gravitational force, F_b is the buoyant force, and F_d is the drag force acting on the particle. Note that the buoyant and drag forces act in the direction opposite to the settling direction.

To simplify our analysis, let's assume the particle is a perfect sphere. We can also confine particle movement to one dimension, where "down" is the positive direction. Since our consideration is limited to movement in one dimension, we can drop the vector notation from forces, velocities, and accelerations.

Using Newton's second law of motion, we can write the net force acting on the particle as the product of the particle mass, m_p, and the acceleration of the particle, a_p. This can be further extended by recognizing that linear acceleration is the time rate of change in linear velocity as shown for the differential and discrete cases in Equation 7.33 and Equation 7.34, respectively. Since the motion of the particle is confined to one dimension, we can drop the vector notation as we've done in the past.

$$\text{Equation 7.33}\quad F_g - F_b - F_d = m_p a_p = m_p \left(\frac{dv_p}{dt} \right)$$

$$\text{Equation 7.34} \quad F_g - F_b - F_d = m_p a_p = m_p \left(\frac{\Delta v_p}{\Delta t} \right)$$

where a_p = acceleration of the settling particle and v_p = velocity of the settling particle.

To further simplify this analysis, we can assume that the particle has reached its terminal velocity, v_t (Equation 7.35).

$$\text{Equation 7.35} \quad \left. \frac{dv_p}{dt} \right|_{v_t} = 0$$

The terminal velocity occurs when a body that moves through a viscous medium stops accelerating. Since the acceleration of the body is zero, the velocity, v_t, is constant. This allows us to set the sum of the forces acting on the particle to zero (Equation 7.36).

$$\text{Equation 7.36} \quad F_g - F_b - F_d = 0$$

Since we've assumed the particle is spherical with a mass, m_p, density, ρ_p, and a diameter, d_p, we can determine the volume of the particle, V_p, using Equation 7.37.

$$\text{Equation 7.37} \quad V_p = \frac{4\pi r_p^3}{3} = \frac{\pi d_p^3}{6}$$

The gravitational force can be written in terms of the particle volume and density as in Equation 7.38.

$$\text{Equation 7.38} \quad F_g = m_p g = V_p \rho_p$$

The buoyant force is proportional to the gravitational force acting on the mass of water that's displaced by the solid spherical particle as shown in Equation 7.39.

$$\text{Equation 7.39} \quad F_b = m_w g = V_p \rho_w g$$

The drag force for a body that's moving through a viscous medium is given in Equation 7.40.

$$\text{Equation 7.40} \quad F_d = \frac{1}{2} C_D A_p \rho_p v_s^2$$

C_D is the drag coefficient and A_p is the cross-sectional area that's perpendicular to the direction of movement. For a sphere, the cross section that's perpendicular to the direction of flow is a circle with a diameter, d_p.

By substituting Equation 7.38, Equation 7.39, and Equation 7.40 into the force balance for a sphere settling at its terminal velocity, we can find the settling velocity, v_s (Equation 7.41) (Majumder 2007).

$$\text{Equation 7.41} \quad v_s = \sqrt{\frac{4g \left(\rho_p - \rho_w \right) d_p}{3 C_D \rho_w}}$$

In the absence of any site-specific data, a particle density of 1,400 kg/m^3 can be used for preliminary calculations.

The drag coefficient for rigid spheres can be calculated according to Equation 7.42. It's important to note that C_D is a function of hydraulic conditions which are characterized using the dimensionless Reynolds number. The form of the Reynolds number for this geometry is shown in Equation 7.43. For this specific situation, Re < 1 for laminar conditions and Re > 1 for turbulent settling. This is different from the case we developed earlier for open channel flow.

$$\text{Equation 7.42} \quad C_D = \frac{24}{Re} + \frac{3}{\sqrt{Re}} + 0.34$$

$$\text{Equation 7.43} \quad Re = \frac{v_s d_p \rho_w}{\mu_w}$$

where μ_w is the absolute viscosity of water (Tchobanoglous and Schroeder 1987).

When spherical particles settle in a viscous medium under laminar flow conditions (Re < 1), the first term in the equation for the drag coefficient dominates and our expression becomes the Stokes equation for settling velocity (Equation 7.44) (Stokes 1851; Tchobanoglous and Schroeder 1987).

$$\text{Equation 7.44} \quad v_{Stokes} = \frac{g\left(\rho_p - \rho_w\right)d_p^2}{18\mu_w}$$

When making calculations for particle settling velocity, it's important to establish the hydraulic flow regime. Under turbulent conditions (Re > 1), Equation 7.41 should be used, while the Stokes equation for the settling velocity applies when Re < 1.

This approach provides a useful framework for analysis but contains several limitations. For example, we've assumed the particles are described as spheres. However, sediment particles aren't perfect spheres and they're not smooth. Rather, they are often faceted with highly irregular surfaces. Likewise, sediment particles in streams and rivers aren't homogeneously sized. Consequently, the approach we've developed provides us with an estimate of the settling behavior of an ideal particle. Think of how real-life conditions might be affected by the assumption that we have discrete, uniform, ideally shaped particles.

7.8.5 DIFFUSION AND DISPERSION

Diffusion is the movement of particles from a region of high concentration to a region of lower concentration. In this case, random molecular motion drives the movement of particles. The diffusion flux, or the movement of particles across a unit area over time, J, is described by Fick's first law. For transport in three dimensions, the diffusion flux is given in Equation 7.45. The negative sign in Fick's first law is included to account for the fact that the movement of mass occurs from regions of high concentration to regions of low concentration.

$$\text{Equation 7.45} \quad J = -D\nabla C\left(x, y, z\right)$$

where D = the diffusion coefficient.

When diffusion occurs in one direction, Equation 7.45 can be simplified. The differential and discrete forms of the one-dimensional diffusion flux equation are presented in Equation 7.46 and Equation 7.47, respectively.

$$\text{Equation 7.46} \quad J = -D\frac{dC}{dx}$$

$$\text{Equation 7.47} \quad J = -D\frac{\Delta C}{\Delta x}$$

Dimensionally, D, is in units of area per unit time (m²/s) and is a function of temperature and fluid viscosity.

Dispersion is the movement of mass from regions of high concentration to regions of lower concentration due to differences in flow velocities along various flow paths. Mixing between flow paths due to diffusion also contributes to mass transport via dispersion.

7.9 A COMBINATION OF TRANSPORT PROCESSES—HOW ADVECTION AND SEDIMENTATION IMPACT THE DOWNSTREAM MOVEMENT OF SOLID PARTICLES

In reality, it's rare that only one transport process is at work in a natural water system. For example, in order to determine how far downstream an impact might occur, we might need to consider both advection and sedimentation.

Example: Assessing the Impact of River Current on the Movement of a Settling Particle

Frequently, we're interested in knowing how long it will take for a particle to be removed through gravity settling. In other cases, we might want to predict how far downstream an impact might occur. In real flowing water systems, sedimentation and advection generally happen simultaneously. To learn more about the impact of advection on particle movement, we'll consider this in two parts—Part A, where particle movement is due only to sedimentation, and Part B, where both sedimentation and advection impact particle movement.

Let's consider a discrete particle with mass, m, radius, r, and density, ρ, that enters the water column at the surface and settles with a velocity, v_s, of 2 x 10⁻⁵ m/s. The water column is 1.5 m deep.

Part A. The first case we'll consider is when there's no stream flow ($v_{stream} = 0$), as shown in Figure 7.7. How long will it take a particle to settle from the top of the water column to the bed? In this case, the trajectory of the particle is straight down. We can find the time needed for the particle to settle by realizing that the settling velocity of the particle is the distance the particle travels, d, over time, t_{settle} (Equation 7.48).

$$\text{Equation 7.48} \quad v_s = \frac{d}{t_{settle}}$$

Part A. Sedimentation only – no advective transport.

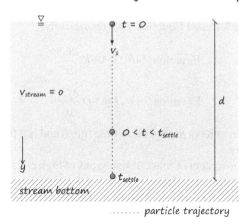

FIGURE 7.7   Case A. Particle movement by sedimentation only.

Since we know d and v_s, we can solve Equation 7.48 for t_{settle} (Equation 7.49).

$$\text{Equation 7.49} \quad t_{settle} = \frac{d}{v_s} = \frac{1.5m}{2\cdot10^{-5}\frac{m}{s}} = 7.5\cdot10^4 s = 20.8hr \cong 21hr$$

In this case, it takes nearly 21 hours for the particle to travel from the water surface to the stream bed. Any particles that are present at depths less than 1.5 m will also be removed from the water column. However, they'll be removed before the particle that started its downward travel at the water surface. Consequently, t_{settle} represents the time needed for 100% removal of particles with a fixed mass, m, radius, r, and density, ρ.

Part B. Now, let's consider the case where the particle settles in the same water column that flows from left to right at 0.05 m/s as shown in Figure 7.8. In this case, the particle will move in both the y and z directions as it settles in the water. We know that it will take $7.5\cdot10^4$ s for the particle to travel from the water surface to the stream bed. However, the particle will no longer settle straight down. Rather, the particle will be displaced further downstream. How far will a particle that enters the water column at the surface take to reach the stream bed? We can find the downstream distance a particle is displaced by relating the stream velocity to the distance and the settling time (Equation 7.50).

$$\text{Equation 7.50} \quad v_{stream} = \frac{z}{t_{settle}}$$

Now, we can solve Equation 7.50 for z (Equation 7.51).

$$\text{Equation 7.51} \quad z = (v_{stream})(t_{settle}) = \left(0.05\frac{m}{s}\right)(7.5\cdot10^4 s) = 3.75\cdot10^3 m = 3.75km$$

Part B. Sedimentation and advective tranport.

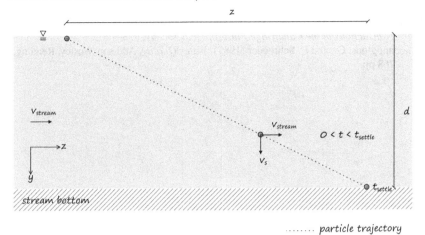

.... particle trajectory

FIGURE 7.8   Case B. Particle movement by sedimentation and advection.

When the stream flows with a velocity of 0.05 m/s, a particle that enters at the water surface the particle will travel 3.75 km downstream before it reaches the stream bottom. If a particle with the same properties entered the water column at a lower depth, it would reach the stream bottom in a shorter distance than a particle that entered at the water surface.

In general, advection will move the particle downstream and sedimentation will move the particle downward toward the stream bed. Something that might not be obvious is the fact that the height at which a particle enters the river will also have an impact on the ultimate downstream distance the particle will travel.

REFERENCES

Arcement, G. (1989). *Guide for selecting Manning's Roughness Coefficients for Natural Channels and Flood Plains*, US Geological Survey Water-Supply Paper 2339, US Geological Survey, Reston, VA, 45 pp.

Bird, R., Stewart, W., and E. Lightfoot (1960). *Transport Phenomena*, 1/e, John Wiley and Sons, New York, 808 pp.

Çengel, Y., and J. Cimbala (2017). *Fluid Mechanics: Fundamentals and Applications*, 4/e, McGraw-Hill, New York, 1056 pp.

Chow, V. (1959). *Open-Channel Hydraulics*, McGraw-Hill, New York, NY, 680 pp.

Clifford, N., Harmar, O., Harvey, G., and G. Petts (2006). "Physical Habitat, Eco-hydraulics and River Design: A Review and Re-evaluation of Some Popular Concepts and Methods," *Aquatic Conservation: Marine & Freshwater Ecosystems*, 16, 389–408.

Haynes, W. (Ed.) (2014). *CRC Handbook of Chemistry and Physics*, 85/e, CRC Press and Taylor and Francis, Boca Raton, FL, 2666 pp.

Majumder, A. (2007). "Settling Velocities of Particulate Systems — a Critical Review of Some Useful Models," *Minerals and Metallurgical Processing*, 24 (4), 237–242.

Manning, R. (1891). "On the Flow of Water in Open Channels and Pipes," *Transactions of the Institution of Civil Engineers of Ireland*, 20, 161–207.

Rouse, H., Siao, T., and S. Nagaratnam (1958). "Turbulence Characteristics of the Hydraulic
      Jump," *Transactions of the American Society of Civil Engineers*, 124 (1), 926–950.
Stokes, G. (1851). "On the Effect of the Internal Friction of Fluids on the Motion of Pendulums,"
      Transactions of the Cambridge Philosophical Society, 9, 8–106.
Tchobanoglous, G., and E. Schroeder (1987). *Water Quality*, Addison-Wesley, Reading, MA,
      768 pp.

8 Alluvial Channels

8.1 ALLUVIAL CHANNELS

Alluvial channels play a key role in defining the relationships between biotic and abiotic components of aquatic and riparian ecosystems. For example, rivers and streams serve as a major conduit for the flow of energy and cycling of materials. While serving as the Earth's circulatory system, rivers and streams transport, mix, concentrate, and transform solids and dissolved constituents including sediment, organic matter, debris, contaminants, microorganisms, and nutrients. Other important ecological services provided by alluvial channels include providing habitats for one or more stages of the lifecycles of aquatic and semiaquatic organisms and supporting the migration of fish, amphibians, microorganisms, and invertebrates.

8.2 STREAM CHARACTERISTICS

Scientists describe and compare streams based on a number of measures that include constancy of flow, stream length, sinuosity, water velocity, water discharge, and stream substrate, among many others.

8.2.1 CLASSIFYING WATER FLOW—CONSTANCY, STEADINESS, AND UNIFORMITY OF FLOW

The set of conditions under which streams have flowing water is known as the constancy of flow. Perennial streams have flowing water year-round because their stream channels are in contact with the groundwater table. Intermittent or seasonal streams have flowing water when the water table is high. Ephemeral stream channels are not in contact with the groundwater table and are fed by surface runoff. Generally, ephemeral streams occur in areas where the groundwater table is very low, such as in highly arid regions.

Stream water flow is also classified based on changes in water velocity over time and/or distance. In steady flow, the time variation in the water velocity through a channel is constant. Unsteady flow occurs when the water velocity changes over time. Under steady flow conditions, the rates of inflow and outflow are constant and equal. This characteristic of water flow will become a central assumption we use when modeling rivers and streams as idealized reactors.

Similarly, the uniformity of flow is related to the change in water velocity with respect to position. In uniform flow, velocity is constant with respect to location in a channel while water velocity varies with position in non-uniform flow. Uniform flow requires a channel with a constant cross section, bed and bank surface roughness, and slope. In non-uniform flow, water velocity and depth vary over distance.

DOI: 10.1201/9781003289630-9

We already know that the water velocity at the stream bed and banks is zero. Likewise, we also know that water levels in streams as well as longitudinal and cross-sectional geometries and bed materials can vary. Consequently, water flow in rivers is typically unsteady and non-uniform. However, when we model these systems, we frequently assume that flow is steady and uniform.

8.3   WATER VELOCITY AND DISCHARGE

In rivers and streams as well as culverts that aren't fully filled, the free water surface is open to the atmosphere. In open channel systems, flow is driven by a difference in elevation between two points located at the free water surface. The slope of this line is known as the hydraulic grade line. In open channel flow, the water surface is parallel to the hydraulic grade line (Sturm 2010).

8.4   ALLUVIAL CHANNEL COMPOSITION AND TRANSPORT MECHANISMS

Alluvial channels are made of loose, unconsolidated sediment that can include clay, silt, sand, and/or gravel. Over time, the shape of alluvial channels is affected by changes in hydraulic conditions that impact the processes of bank and bed erosion, sedimentation, and resuspension. For perspective, photos of a stream reach before (top) and after (bottom) a minor disturbance are presented in Figure 8.1. The log in the stream acts as a trap for fine sediment.

The photo in Figure 8.2 shows how stream flow and runoff work together to remove alluvial material from the stream bank. This reach is also relatively straight with a uniform cross section. In addition to particulate matter, other constituents including nutrients and organic carbon are also mobilized and migrate downstream. Consequently, hydraulic conditions and the composition of bed and bank materials are important factors that impact the overall channel dimensions and morphology. When combined with the effects of vegetation and woody debris, the shape of alluvial channels is constantly evolving. The dynamic nature of alluvial channels makes them home to a diverse range of aquatic and terrestrial inhabitants. In contrast, non-alluvial streams are those channels which are confined by materials that are much less erodible, such as bedrock. Changes in the channel dimensions of non-alluvial streams can take hundreds of years to become apparent.

Like all living things, alluvial channels change in size and shape over time. Initially, a young stream forms a v-shaped valley that splits a broad, flat, landscape. Over time, the valley becomes deeper. At this stage, erosion is not a major factor in channel formation. Young stream channels often contain waterfalls and rapids that are formed or exposed as the channels become deeper. As a stream matures, the valley stops being incised and a floodplain begins to form. With additional time, the valley floor flattens and the floodplain widens. At this point, sediment begins to accumulate and erosion becomes an important factor in the future dimensions of the channel. Additionally, the flat channel bottom will begin to meander.

"Bankfull depth" means the average vertical distance between the channel bed and the estimated water surface elevation required to completely fill the channel to a

FIGURE 8.1 A reach of Forman Creek (Knox County, IL, USA) before (top) and after (bottom) a minor disturbance. Photos by Robin Bauerly.

FIGURE 8.2   A cutting bank on Forman Creek (Knox County, IL, USA). During high water events, runoff and stream flow work together to scour material from the unprotected alluvial material that makes up the stream bank. Photo by Robin Bauerly.

point above which water would enter the floodplain or intersect a terrace or hillslope (Huggett 2016).

8.5   SINUOSITY

Sinuosity is a stream's tendency to move back and forth in an "S pattern" across its floodplain over time. Geomorphologists calculate the sinuosity of a stream reach (Equation 8.1) as the ratio of the length of the valley created by the stream, L_v, divided by the straight-line distance between the beginning and end of a stream reach, L_s, as presented in Figure 8.3.

$$\text{Equation 8.1}\quad S = L_v/L_s$$

where S = sinuosity (dimensionless), L_v = length of the valley created by the stream, and L_s = straight-line distance between the beginning and end of a stream reach.

The valley length is found by following the stream thalweg (the longitudinal trace of the deepest part of reach from its source to its terminus). In this context, a straight stream would have a sinuosity of 1. As sinuosity increases, the stream reach will take

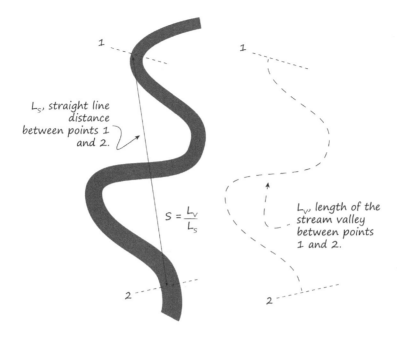

FIGURE 8.3   Sinuosity of a stream reach between points 1 and 2 (Huggett 2016).

on a more tortuous, less regular meandering path. In practice, sinuosity values from 1 to 3 are typical. Meandering channels have a sinuosity index value exceeding 1.5 (Wilzbach and Cummins 2019). Alternatively, sinuosity may be determined as the ratio of the straight-line slope to the valley slope. However, making sinuosity calculations based on stream slopes can be of limited utility in areas with relatively little change in topographic relief.

8.6   ALLUVIAL CHANNEL COMPOSITION AND TRANSPORT MECHANISMS

8.6.1   BED MATERIAL

In addition to sinuosity, alluvial channels are classified by the characteristics of their bed material—this includes both dissolved and solid materials. In a manner similar to the way we characterized the concentration of solids in water, the total load carried by alluvial streams includes solid and dissolved fractions (Figure 8.4). Among the solid materials, some are suspended in the water column while others reside on the stream bed. In general, particles that make up bedload are larger and/or denser than those that contribute to suspended loading (Huggett 2016).

Bedload is the material that's carried by a stream, usually by rolling along or bouncing up and down on the channel bed. The distribution of solid particles between the suspended and bed/deposited load depends on factors that include the size and density of particles and the water velocity. In bedload dominated channels,

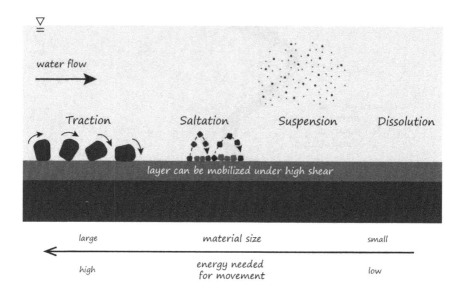

FIGURE 8.4   Schematic of bed material transport in an alluvial channel (Huggett 2016).

more than 10% of the total load is made up of heavier particles that move along the stream bed. In these streams, larger/more dense particles are moved by a process of traction (rolling along the stream bottom). Particles that are smaller/less dense can be lofted into the water column, but then settle back to the stream bed through a process known as saltation.

The movement of bedload materials can be continuous or intermittent depending on the water flow conditions. In general, rivers with higher discharge (flow rate) can carry a higher bed load than those with lower discharge. Consequently, flooding events often result in large-scale transport of bedload material. When a stream's ability to carry bedload is exceeded, the stream channel becomes obstructed. This results in the formation of braided channels. In this regard, any factor that impacts the ability of a stream to move bedload materials—including the presence of vegetation and woody debris—can affect the cross-sectional dimensions and shape of a bedload dominated channel.

In suspended-load channels, less than ~3% of the total load is bedload. Suspended particulate matter is usually made up of fine particles clays and fine silt. When suspended in water, the electrostatic interactions between particles and water molecules help to keep the particles in the water column. This is typical of the Mississippi River, where the water clarity can often be adversely affected by the high concentration of particles suspended in the water column. This is a good time to remember that discrete particle settling can be described using Stokes' law. However, it's also important to recall that in most natural systems, the size and material properties of particles are rarely homogeneous. As a result, we need to consider limiting conditions when we look to predict the movement of particles in quiescent and flowing water systems.

8.6.2 HYPORHEIC ZONE

Another feature of alluvial channels is the development of a hyporheic zone, where water moves between the bulk flow and the interstitial spaces between alluvial deposits. This movement of water plays a key role in the exchange of gases, solutes, and microorganisms. Hyporheic flow can include a combination of surface water and groundwater. When compared to flow in the main channel, the residence time of water in the hyporheic zone is longer. This has direct impacts on a channel's ability to process carbon and nutrients. In some cases, the retention, transformation, and subsequent release of constituents from the hyporheic zone can reduce the rate at which anticipated processes are seen in downstream waters (Grimm and Fisher 1984). The rate that water moves between the bulk flow and the hyporheic zone is related to the properties of the stream bed materials (Bencala 2000; Findlay 1995).

8.7 AQUATIC ECOSYSTEMS AND HABITAT—ECOLOGICAL AND GEOMORPHOLOGICAL PERSPECTIVES

As you may recall from Chapter 1, Odum (1964) defined the ecosystem as the basic functional unit of organisms and their environment. This includes the way organisms interact among themselves and with all other biotic and abiotic constituents in a system. In this regard, scale and size are less important than the interactions between biotic and abiotic parts of an environmental system. So, an ecosystem can range in size from an entire watershed to a pond, or a unit volume of a water-filled stream bed. This might bring to mind the concept of the control volume used when performing material balances.

In general, the term habitat is used to indicate the area, resources, and abiotic constituents that are used by an organism or group of organisms. In flowing water systems, both ecological and geomorphological approaches have been used to describe habitats. This is another example of how aquatic environmental scientists and engineers can benefit from being fluent in a wide range of disciplines. Using the ecological approach, habitats are defined based on the biotic and abiotic factors that support distinct assemblages of aquatic—and riparian—organisms (Harper *et al.* 1992). Consequently, it's important to recognize that when we refer to an environmental compartment as a habitat, we're necessarily including factors that go beyond a particular place and extend to organisms and their relationships with other living and abiotic factors. In contrast, the geomorphological approach to defining habitat is built around the physical conditions that allow us to distinguish the living places of different organisms or groups of organisms. For example, in alluvial streams, riffles, runs, pools, and glides are distinguished by differences in physical characteristics such as flow velocity, water depth, and the characteristics of bed and bank materials (Wadeson 1994).

A similar divide is seen in the different approaches to describing habitats. For example, using the geomorphological approach, stream habitats are identified based on geomorphological features that are distinguished by differences in parameters such as flow and water column depth. Both approaches share a number of common elements and neither approach is inherently better than the other. Rather, it's important to have a working understanding of both to understand and integrate the perspectives provided by both approaches to make informed decisions. For example, consider what happens when a beaver builds a dam. When water is impounded, the extent of open water increases, which can increase the rate at which water is lost by evaporation. On the other hand, the presence of standing water can help to recharge local groundwater supplies and downstream channels can be converted from perennial to ephemeral flows (Pollock *et al.* 2003).

Each of these physical changes has greater reaching impacts that can affect both biotic and abiotic parts of an ecosystem. For example, a beaver impoundment can increase local sediment and nutrient storage. Other changes include the development of a standing water (lentic) ecosystem in a formerly flowing water (lotic) system. As more water is held upstream from the beaver impoundment, streambeds can become dominated by anaerobic conditions which can be impactful to the food web (Pollock *et al.* 2003). Similarly, networks of burrows created by crayfish have been found to impact the geomorphology of streams. By providing paths for water to infiltrate stream banks, erosion can become more significant (Statzner *et al.* 2000).

8.8   HABITATS IN ALLUVIAL CHANNELS

A typical alluvial stream consists of three major habitats that are found in a sequential pattern of riffles, runs, and pools (Figure 8.5). Some scientists and engineers refer to the segment of a stream that connects a pool to a subsequent riffle as a glide. In this case, the sequence of habitats would be riffle, run, pool, and glide. Each habitat type is distinguished based on its hydrologic and geomorphological characteristics.

Riffles are shallow stream segments where water flows through course-bedded materials. Turbulence is typically higher in riffles than in runs or pools. Riffles serve as a prime habitat for many macroinvertebrate species including *Ephemeroptera* (mayflies), *Plecoptera* (stoneflies), and *Trichoptera* (caddisflies) which are widely used when conducting rapid bioassessment protocols to characterize stream health (Barbour *et al.* 1999). Consequently, many fish spawn and feed in riffles. Additionally, when water temperatures are elevated, riffle habitats often have higher dissolved oxygen concentrations than other more quiescent habitat types. Consequently, fish often inhabit riffles to scavenge for food and to alleviate low oxygen conditions that are experienced frequently in summer (Barbour *et al.* 1999; Matthaei *et al.* 1999, 2000). The combination of high and low velocity zones helps maintain a diverse biological assemblage in alluvial channels (Jenkins *et al.* 1984; Tockner *et al.* 2003).

Runs are stream reaches with laminar flow and a more or less uniform stream cross-section. Typically, the water level in runs is deeper than in riffle habitats. They also have relatively good light penetration, which helps to support the photosynthetic organisms that are important to the food chain. Runs (or glides) often connect the outlet of pools and the start of riffles. They are frequently a home to smaller fish that are unable to compete for resources in pools. When not protected by cover, the water temperature in relatively shallow runs can rise during the day. In some cases, this can be a significant source of stress for aquatic receptors.

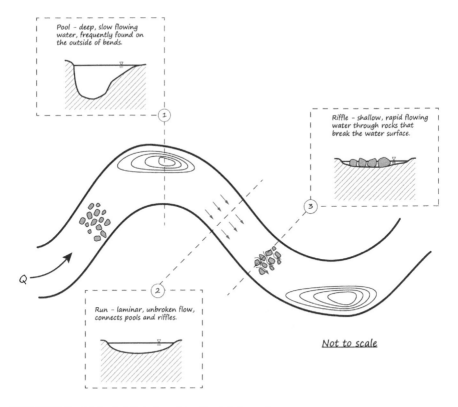

Pool – deep, slow flowing water, frequently found on the outside of bends.

Riffle – shallow, rapid flowing water through rocks that break the water surface.

Run – laminar, unbroken flow, connects pools and riffles.

Not to scale

FIGURE 8.5   Major habitats found in alluvial channels.

A pool is a stream segment that's deeper than average with a slower than average water velocity. Pools are often found on the outside bend in a meandering channel. At the same time, deposits of alluvium on the inner bend of a meander can form a point bar (see Figure 8.6). These features are the result of the asymmetric deposition of stream energy on the banks and bed when a stream is forced to change course on a meander. Pools often serve as a refuge for fish when water levels in the channel are low (Negishi *et al.* 2002). Pools are also a collection zone for biological drift, which can be a significant source of food for aquatic organisms. Other habitat types/hydraulic features—including backwaters and eddies—can also serve a similar purpose.

You'll note that the cross section of a pool habitat in Figure 8.5 is much different from the idealized cross sections we saw previously (Figure 7.2) where we assumed centerline symmetry. In alluvial channels, any imbalance in the energy of flowing water is reflected in changes to the dimensions of the stream bed and channel. These are known as compound cross sections and can often describe the overall geometry of as a combination of simpler shapes.

FIGURE 8.6 A point bar of deposited material is commonly found on the inner bend of a stream meander (Franklin Creek, Ogle County, IL, US. Photo by Roger C. Viadero, Jr., May 8, 2018).

Example: Calculating Channel Dimensions for a Compound Cross Section

Calculate the hydraulic radius, R_h, of the compound stream cross section presented in Figure 8.7, which is representative of a pool habitat. To determine R_h, we

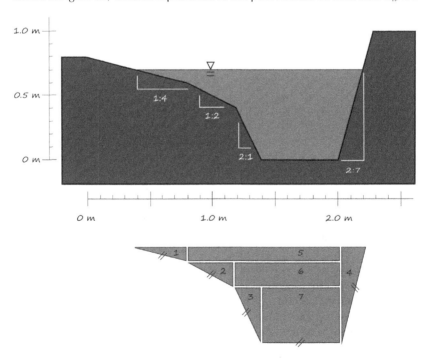

FIGURE 8.7 A compound stream cross section broken into simpler shapes to calculate cross-sectional area, wetted perimeter, and hydraulic radius. Only the hatched line segments contribute to the wetted perimeter.

TABLE 8.1
Areas and Wetted Perimeters Needed to Calculate the Hydraulic Radius for the Compound Cross Section.

Section	Area, m²	p_w, m	R_h, m
1	0.02	0.41	0.049
2	0.08	0.45	0.178
3	0.08	0.45	0.178
4	0.14	0.73	0.311
5	0.06	–	–
6	0.08	–	–
7	0.24	0.60	0.400
		Sum	1.107

need to know the channel's cross-sectional area, A, and the wetted perimeter, p_w. To do this, we can break the cross section into seven simpler shapes, as shown in Figure 8.7. The area of each section can be determined using basic relationships for the areas of triangles and rectangles. It's important to recall that only the line segments of the triangles and rectangles that describe the larger compound cross section contribute to the wetted perimeter. Areas and wetted perimeters for the seven segments are presented in Table 8.1. Based on these data, the hydraulic radius for the compound cross section is 1.12 m.

Previously, the role of vegetation and woody debris were mentioned as stream features that can alter the ability of a channel to move bedload. While it might not be obvious at first, woody riparian vegetation is the major source of woody debris in alluvial channels. Riparian vegetation also helps to moderate water temperatures by providing shade. Additionally, riparian vegetation provides leaf litter which serves as a source of food for organisms in the water column, stabilizes stream banks, and adds to habitat complexity (Sweeney 1992; Sweeney *et al.* 2004; Trimble 1997).

8.9   WATER DEPTH AND VELOCITY AS INDICATORS OF HABITAT TYPES

An understanding of the relationship between water velocity and depth can provide valuable insight into conditions which might favor the suspension of solids in the water column and/or those which result in increased stream bank and bed erosion. However, the Froude number, Fr, is also a valuable tool that can be used to predict the assemblages of aquatic organisms that are likely to be found in different habitats. For example, Hilldale and Mooney (2007) developed specifications for aquatic habitats in streams as a function of Fr; where pools occur at Fr < 0.09, riffles at Fr > 0.41, and runs at 0.09 < Fr < 0.41. Alternatively, Jowett characterized pools as stream reaches where Fr < 0.18 (Jowett 1993). It's worth noting that the Froude number in each case describes subcritical flow. In this case, the water velocity and/or depth are below critical values. Consequently, when flow disturbances are created, part of the resulting waveform can be transmitted upstream.

Example: Assessing the Favorability of Stream Habitats Based on Hydraulic Factors

You're the leader of a stream restoration team. In this work, you will need to create pools, riffles, and runs along a stream reach. If the average stream discharge is 1.5 m³/s and the maximum channel width is 8 m, determine the stream habitats that are likely to form at each of five velocities (0.2, 0.4, 0.6, 0.8, 1.0 m/s) that are characteristic of water flow in the channel. Over this range of velocities, have all habitat types been represented? If not, suggest other conditions that should be investigated.

We're given a fixed flowrate and one channel dimension and need to determine habitat conditions for a range of water velocities. The only relationship we know that will allow us to relate channel dimensions and water velocity to habitat type is the Froude Number (Equation 7.20). To determine Fr, we need to know the water velocity and depth. We can use the flowrate and width to find the depth of the water column by using the relationship between flowrate, velocity, and cross-sectional area. In the absence of any other information, let's assume the channel has a rectangular cross-sectional geometry. This allows us to write an expression to calculate the depth (Equation 8.2).

$$\text{Equation 8.2} \quad d = \frac{Q}{vw} = \frac{1.5 \frac{m^3}{s}}{v(8m)} = \frac{0.1875}{v}$$

The water depth and Froude Number for each water velocity is presented in Table 8.2. Based on the criteria suggested by Hilldale and Mooney (2007), the corresponding habitat types are all identified as pools, runs, and riffles.

TABLE 8.2

Calculations Used to Predict Habitat Type in a Stream With a Fixed Width and Constant Flowrate.

v, m/s	d, m	Fr	Habitat
0.2	0.94	0.07	Pool
0.4	0.47	0.19	Run
0.6	0.31	0.34	Run
0.8	0.23	0.53	Riffle
1.0	0.19	0.74	Riffle

8.10 STRAHLER NUMBER—BRINGING ORDER TO STREAM CLASSIFICATION

In the early 1950s, Professor Arthur Strahler proposed a system to classify streams based on their order (Strahler 1952). Strahler's work was built on an earlier study by Robert Horton (Horton 1945). Consequently, stream order is often referred to as the "Strahler Number" or "Strahler-Horton Number." Using this approach, streams

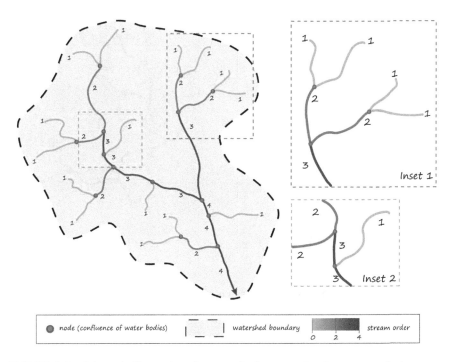

FIGURE 8.8   Schematic illustration of a network of streams that form a watershed.

are idealized as a tree-like framework, as presented in Figure 8.8. Each branch on the tree is referred to as a "reach," which is simply a length of stream between two points. While "between two points" is admittedly vague, typically they are locations of significance, such as a stream confluence, a gaging station, or a natural feature, *etc.*

Let's consider a system of streams where the confluences of waterbodies are known as "nodes." The smallest reaches are classified as "first order." The confluence of two first-order streams is a second-order stream. Similarly, the confluence of two second-order streams is a third-order stream as shown in Figure 8.9, inset 1, and so on. In the mass balance section, we saw how the Ohio River (an eighth-order channel) is formed from the confluence of the Allegheny and Monongahela Rivers, which are both seventh-order streams. Similarly, the Mississippi River is sixth order through its upper reach and becomes tenth order below the confluence with the Ohio River.

When two streams of different order combine, there's no change in the order of the stream (Figure 8.9, Inset 2). Using this approach provides a measure of the size of a reach relative to other streams in the network.

One case in which assigning an order to streams is a little more complicated is a system of reaches that includes a large body of water (Figure 8.9). In this case, stream order is determined by working from the top of the watershed down—not around the edges of the lake. For example, the stream that enters the lake is second

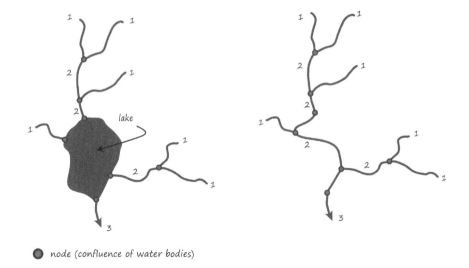

● node (confluence of water bodies)

FIGURE 8.9 Determining stream order in a network of streams that contains a standing body of water.

TABLE 8.3
Select Rivers and Their Highest
Respective Orders (USGS 2023).

River	Order
Ohio River	8
Lower Colorado River	9
Mississippi River	10
Amazon River	12

---

order. Working our way down, the first-order stream from the left doesn't change the order of the "main channel." However, after the confluence of the second order stream from the right, the stream that drains the lake becomes third order.

First- through third-order streams are known as headwater streams. Rivers are classified as streams of the sixth order and higher. Several rivers and their respective orders are presented in Table 8.3. In comparison to headwater streams, the channel slopes of rivers are lower and have slower moving water. However, the conveyance of rivers is much greater than that of headwater streams.

8.11   HEADWATER STREAMS

While headwater streams might be small, they actually make outsized contributions to the overall length of streams and offer unique ecological services while also

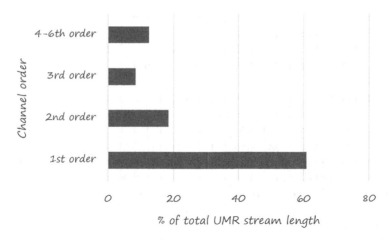

FIGURE 8.10 Distribution of Upper Mississippi River streams as a function of Strahler stream order (USGS 2023).

providing economic dividends. With an area of 3.2 million km² (2.4 million mi²), the Mississippi River Basin covers around 40% of the continental United States and about 12.5% of North America. The rivers that make up the Mississippi River Basin drain through 31 US states and two Canadian provinces (US EPA 2023).

The Upper Mississippi River Basin includes the headwaters of the Mississippi River at Lake Itasca, MN, and flows through five states. The Upper Mississippi River (52,072 km²; 20,105 mi²) is impounded by 29 lock and dam structures. Shortly after its confluence with the Missouri and Ohio Rivers, the Mississippi flows into the Lower Basin, where it's no longer restricted by locks and dams. People frequently associate all parts of the Mississippi River Basin with the huge main channel. However, as we'll see, the Mississippi River Basin and other large drainage basins across the globe are actually dominated by low-order streams. For example, in the Upper Mississippi River Basin, 87.6% of stream length is made up of headwater streams (Figure 8.10) (USGS 2023). In contrast, only 12.4% of stream channels in the Upper Mississippi River Basin are larger river channels—fourth-, fifth-, and sixth-order streams. The high percentage of headwater streams is not unique to the Upper Mississippi River Basin. For example, headwater streams have been reported to account for 60–80% of the total stream length within a catchment (Schumm 1956; Shreve 1969; Colvin *et al.* 2019).

Since headwater streams account for such a large fraction of the total stream length, their role in mitigating the downstream impacts of flooding can be significant. These dense networks of small streams intercept precipitation and keep it from reaching larger waterbodies. They also help to dissipate stream energy as flowing water travels a meandering path to lower elevations.

Now, reconsider the role of other in-channel materials—including cobbles and woody debris—and think of the amount of energy that's dissipated in collisions with flowing water. Headwater streams play an equally important role when water is in short supply. In the hyporheic zone, surface and shallow groundwater can mix. In fact, the USGS estimated that almost half of the surface water in streams and rivers actually comes from groundwater (Alley *et al.* 2019). The hyporheic zone in alluvial channels also serves as the conduit that allows wetlands to play a key role as headwaters.

In addition to regulating the flow of water, headwater streams also help to maintain the health of streams. For example, the ability of stable streams to assimilate and manage sediment is known to help mitigate downstream solid loading. As a consequence, more light is able to penetrate the water column and support primary productivity. Since sediment often contains a variety of constituents including organic carbon and nutrients, the effective management of solids can also aid in the management of eutrophication due to high phosphorus concentrations. This, in turn, helps to reduce the depletion of oxygen that can result from the rapid overgrowth of algae (Dodds 2006). Likewise, the bacteria and fungi that inhabit headwater streams are capable of mitigating the impacts of inorganic nitrogen loading (Peterson *et al.* 2001).

Headwater streams are also a significant economic resource. Any doubt of this fact can be assuaged by looking at the price of fishing gear and the willingness of anglers to travel far and wide to stalk their quarry. Nadeau and Rains (2007) estimated that headwater streams in the conterminous US and Hawaii provided ecosystem services valued at approximately \$15.7 trillion/yr (~\$22.2 trillion/yr in 2022). Lane and D'Amico (2016) extended this analysis and reported that the wetlands adjacent to the floodplain added \$673 billion/yr (~\$820 billion/yr in 2022) in ecosystem services. While many economists attempt to quantify the value provided by headwater streams, Colvin *et al.* (2019) offered a much more precautionary approach to their valuation. As opposed to viewing headwater streams as an economic asset, they suggested considering the negative consequences—both ecological and economic—of failing to preserve and maintain the abundance and integrity of headwater streams. This brings to mind the saying, "an ounce of prevention is worth more than a pound of cure."

When using common systems of stream classification, headwater streams are often cited as having relatively steep slopes and correspondingly fast flowing water. It's important to place these general classifications in proper context. For example, according to the Rosgen stream classification system that serves as a standard method when classifying streams and rivers according to their fluvial geomorphological characteristics, headwater streams often have steep slopes (> 4%) (Rosgen 1994). This certainly applies for the areas where the method was developed and for other locales where it has been adapted for local use, including more mountainous regions in the eastern US. However, this particular classification system should be applied with caution in areas where there is little topographic relief, including much of the middle of the US.

8.12   CLIMATE CHANGE AND BIOLOGICAL IMPACTS

We've spent a considerable amount of time qualifying and quantifying the characteristics of water, describing flow and transport in waterways, and explaining habitat features in alluvial channels. Given the increase in the frequency and intensity of extreme environmental conditions, it's timely to consider how these changes can impact aquatic environmental systems. It's also a good time to recall the structural stream framework (Figure 1.1) (Harman *et al.* 2012; Fischenich 2006) that allowed us to relate ecosystem structure and function.

Consider the increased frequency and severity of flooding and the impact(s) this has on the functions that support unique habitats and assemblages of biota. Starting at the bottom of the structural stream framework, flooding has clear implications on the abundance of both surface and ground water in a drainage basin. However, a net change in watershed hydrology can affect primary productivity by changing water clarity and nutrient availability (Lindholm *et al.* 2007). Initially, primary productivity might be inhibited by high water levels. However, when flood waters recede, productivity often increases in response to the abundance of nutrients that are deposited by flood waters (Paerl *et al.* 2011). After large flooding events, excessive primary productivity can also occur. For example, a record sized algal bloom in Lake Erie was attributed to an influx of phosphorus from heavy rainfall (King *et al.* 2017). In other cases, there can be a shift in the composition of a biological community. After flooding in the Lake Winnipeg Basin, there was a shift in the phytoplankton community to include a greater fraction of cyanobacteria (McCullough *et al.* 2012).

As flood waters rise, the hydraulic characteristics of streams and rivers change. For example, rapidly moving flood waters can mobilize sediment particles that are typically part of bedload. This can include substrates that are home to benthic macroinvertebrates. This is reflected by a reduction in biodiversity and species density (Fornaroli *et al.* 2019). When rivers and streams can no longer accommodate flood waters, the riparian zone becomes inundated. In alluvial channels, flood waters reshape the dimensions, pattern, and profile of streams. When combined with the movement of stream substrate and bed and bank erosion, existing habitat can be altered while new habitat is created. The presence of woody debris that's entrained in floodwaters can also cause changes in channel geometry and sediment composition (Danehy *et al.* 2012).

Gholizadeh (2021) found that the impact of flooding on benthic macroinvertebrate communities was comparable to those seen in heavily polluted waters. While the benthic macroinvertebrate communities were able to recover from flooding, macroinvertebrate richness, distribution, and community composition were adversely impacted. In this case, the impacts of flooding on hydrology, hydraulics, geomorphology, and physicochemical characteristics and processes were so significant that lasting changes to the biological community occurred. Junk *et al.* (1989) found that the increased frequency and extent of flooding has reduced the resiliency of aquatic systems. This fact provides ample support for the need to better understand the relationships between aquatic environmental compartments.

REFERENCES

Alley, W., Reilly, T., and O. Franke (1999). *Sustainability of Ground-Water Resources*, US Geological Survey Circular 1186, US Geological Survey, Denver, CO.

Barbour, M., Gerritsen, J., Snyder, B., and J. Stribling (1999). *Rapid Bioassessment Protocols for Use in Streams and Wadeable Rivers: Periphyton, Benthic Macroinvertebrates and Fish*, 2nd ed., EPA 841-B-99–002, US Environmental Protection Agency, Office of Water, Washington, DC.

Bencala, K. (2000). "Hyporheic Zone Hydrological Processes," *Hydrological Processes*, 14 (15), 2797–2798.

Colvin, S., Sullivan, S., Shirey, P., Colvin, R., Winemiller, K., Hughes, R., Fausch, K., Infante, D., Olden, J., Bestgen, K., Danehy, R., and L. Eby (2019). "Headwater Streams and Wetlands are Critical for Sustaining Fish, Fisheries, and Ecosystem Services," *Fisheries*, 44 (2), 73–91.

Danehy, R., Bilby, R., Langshaw, R., Evans, D., Turner, T., Floyd, W., Schoenholtz, S., and S. Duke (2012). "Biological and Water Quality Responses to Hydrologic Disturbances in Third-Order Forested Streams," *Ecohydrology*, 5, 90–98.

Dodds, W. (2006). "Eutrophication and Trophic State in Rivers and Streams," *Limnology and Oceanography*, 51, 671–680.

Findlay, S. (1995). "Importance of Surface-Subsurface Exchange in Stream Ecosystems: The Hyporheic Zone," *Limnology and Oceanography*, 40 (1), 159–164.

Fornaroli, R., Calabrese, S., Marazzi, F., Zaupa, S., and V. Mezzanotte (2019). "The Influence of Multiple Controls on Structural and Functional Characteristics of Macroinvertebrate Community in a Regulated Alpine River," *Ecohydrology*, 12 (2), https://doi.org/10.1002/eco.2069.

Gholizadeh, M. (2021). "Effects of Floods on Macroinvertebrate Communities in the Zarin Gol River of Northern Iran: Implications for Water Quality Monitoring and Biological Assessment," *Ecological Process*, 10 (46), https://doi.org/10.1186/s13717-021-00318-0.

Grimm, N., and S. Fisher (1984). "Exchange Between Interstitial and Surface Water: Implications for Stream Metabolism and Nutrient Cycling," *Hydrobiologia*, 111 (3), 219–228.

Harman, W., Starr, R., Carter, M., Tweedy, K., Clemmons, M., Suggs, K., and C. Miller (2012). *A Function-Based Framework for Stream Assessment and Restoration Projects*, US Environmental Protection Agency, Office of Wetlands, Oceans, and Watersheds, Washington, DC, EPA 843-K-12–006.

Harper, D., Smith, C., and P. Barham (1992). "Habitats as the Building Blocks for River Conservation Assessment," in Boon, P., Callow, P., and G. Petts (Eds.), *River Conservation and Management*, Wiley and Sons, New York, 311–319 pp.

Hilldale, R., and D. Mooney (2007). *Identifying Stream Habitat Features with a Two-Dimensional Hydraulic Model*, Technical Series No. TS-YSS-12, US Bureau of Reclamation Technical Service Center, Denver, CO. 33 pp., www.usbr.gov/pn/programs/storage_study/reports/ts-yss-12/2Dmodel.pdf.

Horton, R. (1945). "Erosional Development of Streams and Their Drainage Basins: Hydro-Physical Approach to Quantitative Morphology," *Geological Society of America Bulletin*, 56 (3), 275–370.

Huggett, R. (2016). *Fundamentals of Geomorphology*, 4/e, Routledge, London, 578 pp.

Jenkins, R., Wade, K., and E. Pugh (1984). "Macroinvertebrate-Habitat Relationships in the River Teifi Catchment and the Significance to Conservation," *Freshwater Biology*, 14, 23–42.

Jowett, I. (1993). "A Method for Identifying Pool, Run, and Riffle Habitats from Physical Measurements," *New Zealand Journal of Marine and Freshwater Research*, 27, 242–248.

Junk, W., Bayley, P., and R. Sparks (1989). "The Flood Pulse Concept in River-Floodplain Systems," *Canadian Special Publication of Fisheries and Aquatic Sciences*, 106, 110–127.

Kaufmann P., Levine, P., Robinson, E., Seeliger, C., and D. Peck (1999). *Quantifying Physical Habitat in Wadeable Streams*, US Environmental Protection Agency, Washington, DC. EPA/620/R-99/003.

King, K., Williams, M., Johnson, L., Smith, D., LaBarge, G., and N. Fausey (2017). "Phosphorus Availability in Western Lake Erie Basin Drainage Waters: Legacy Evidence Across Spatial Scales," *Journal of Environmental Quality*, 45, 466–469.

Lane, C., and E. D'Amico (2016). "Identification of Putative Geographically Isolated Wetlands of the Conterminous United States," *Journal of the American Water Resources Association*, 52, 705–722.

Matthaei, C., Arbuckle, C., and C. Townsend (1999). "Patchy Surface Stone Movement During Disturbance in a New Zealand Stream and its Potential Significance for the Fauna," *Limnology and Oceanography*, 44, 1091–1102.

Matthaei, C., Arbuckle, C., and C. Townsend (2000). "Stable Surface Stones as Refugia for Invertebrates During Disturbance in a New Zealand Stream," *Journal of the North American Benthological Society*, 19, 82–93.

McCullough, G., Page, S., Hesslein, R., Stainton, M., Kling, H., Salki, A., and D. Barber (2012). "Hydrological Forcing of a Recent Trophic Surge in Lake Winnipeg," *Journal of Great Lakes Research*, 38, 95–105.

Nadeau, T., and M. Rains (2007). Hydrological Connectivity Between Headwater Streams and Downstream Waters: How Science Can Inform Policy," *Journal of the American Water Resources Association*, 43 (1), 118–133, doi:10.1111/j.1752-1688.2007.00010.x.

Negishi, J., Inoue, M., and M. Nunokawa (2002). "Effects of Channelization on Stream Habitat in Relation to a Spate and Flow Refugia for Macroinvertebrates in Northern Japan," *Freshwater Biology*, 47, 1515–1529.

Odum, E. (1964). "The New Ecology," *BioScience*, 14, 14–16.

Paerl, H., Hall, N., and E. Calandrino (2011). "Controlling Harmful Cyanobacterial Blooms in a World Experiencing Anthropogenic and Climatic Induced Change," *Science of the Total Environment*, 409, 1739–1745.

Peterson, B., Wollheim, W., Mulholland, P., Webster, J., Meyer, J., Tank, J., Marti, E., Bowden, W., Valett, H., Hershey, A., McDowell, W., Dodds, W., Hamilton, S., Gregory, S., and D. Morrall (2001). "Control of Nitrogen Export from Watersheds by Headwater Streams," *Science*, 292 (5514), 86–90, doi:10.1126/science.1056874.

Pierson, S., Rosenbaum, B., McKay, L., and T. Dewald. (2008). "Strahler Stream Order and Strahler Calculator Values in NHDPlus," US Geological Survey, ftp://ftp.horizon-systems.com/NHDPlusExtensions/SOSC/SOSC_technical_paper.pdf

Pollock, M., Heim, M., and D. Werner (2003). "Hydrologic and Geomorphic Effects of Beaver Dams and Their Influence on Influence on Fishes," *American Fisheries Society Symposium*, 37, 213–233.

Rosgen, D. (1994). "A Classification of Natural Rivers," *Catena*, 22, 169–199.

Schumm, S. (1956). "Evolution of Drainage Systems and Slopes in Badlands at Perth Amboy, New Jersey," *Bulletin of the Geological Society of America*, 67, 597–646.

Shreve, R. (1969). "Stream Lengths and Basin Areas in Topologically Random Channel Networks," *Journal of Geology*, 77, 397–414.

Statzner, B., Fièvet, E., Champagne, J., Morel, R., and E. Herouin (2000). "Crayfish as Geomorphic Agents and Ecosystem Engineers: Biological Behavior Affects Sand and Gravel Erosion in Experimental Streams," *Limnology and Oceanography*, 45 (5), 1030–1040.

Strahler, A. N. (1952), "Hypsometric (Area-Altitude) Analysis of Erosional Topology," *Geological Society of America Bulletin*, 63 (11), 1117–1142.

Sturm, T. (2010). *Open Channel Hydraulics*, 2/e. McGraw Hill, New York, 546 pp.

Sweeney, B. (1992). "Streamside Forests and the Physical, Chemical, and Trophic Characteristics of Piedmont Streams in Eastern North America," *Water Science and Technology*, 26, 2653–2673.

Sweeney, B., Bott, T., Jackson, J., Kaplan, L., Newbold, J., Standley, L., Hession, W., and R. Horwitz (2004). "Riparian Deforestation, Stream Narrowing, and Loss of Stream Ecosystem Services," *Proceedings of the National Academy of Sciences, USA*, 101, 14132–14137.

Talbot, C., Bennett, E., Cassell, K., Hanes, D., Minor, E., Paerl, H., Raymond, P., Vargas, R., Vidon, P., Wollheim, W., and M. Xenopoulos (2018). "The Impact of Flooding on Aquatic Ecosystem Services," *Biogeochemistry*, 141, 439–461, https://doi.org/10.1007/s10533-018-0449-7.

Tockner, K., Ward, J., Arscott, D., Edwards, P., Kollmann, J., Gurnell, A., Petts, G., and B. Maiolini (2003). "The Tagliamento River: A Model Ecosystem of European Importance," *Aquatic Sciences*, 65, 239–253.

Trimble, S. (1997). "Stream Channel Erosion and Change Resulting from Riparian Forests," *Geology*, 25, 467–469.

US EPA (2023). *The Mississippi/Atchafalaya River Basin (MARB)*, Mississippi River/Gulf of Mexico Hypoxia Task Force, www.epa.gov/ms-htf/mississippiatchafalaya-river-basin-marb, accessed March 21, 2023.

USGS (2023). "National Hydraulic Database Plus, High Resolution," www.usgs.gov/national-hydrography/nhdplus-high-resolution.

Wadeson, R. (1994). "A Geomorphological Approach to the Identification and Classification of Instream Flow Environments," *Southern African Journal of Aquatic Science*, 20 (1–2), 38–61.

Wilzbach, M., and K. Cummins (2019). "Rivers and Streams: Physical Setting and Adapted Biota," in Fath, B. (Ed.), *Encyclopedia of Ecology*, 2/e, Elsevier Science and Technology, Amsterdam, 594–606 pp.

ADDITIONAL READING

US Environmental Protection Agency (USEPA) (2015). *Connectivity of Streams and Wetlands to Downstream Waters: A Review and Synthesis of the Scientific Evidence (Final Report)*, EPA/600/R-14/475F, US Environmental Protection Agency, Washington, DC, https://cfpub.epa.gov/ncea/risk/recordisplay.cfm?deid=296414.

9 Ideal Reactors

9.1 IDEAL REACTORS

A reactor is a system that's operated under controlled conditions. Constituents are conveyed into the reactor through one or more inlet flows. They reside in the reactor for a period equal to the reactor's hydraulic retention time. During this period, the constituents can be transformed through physical, chemical, and biological interactions before being discharged from the reactor. Fluid dynamics and chemical kinetics are the two elements that govern the behavior and performance of a reactor. Four types of reactor systems/configurations will form the foundation for our analysis of water systems; these include batch reactors, continuous flow stirred tank reactors (CFSTR), CFSTRs in series, and plug flow reactors (PFRs) (Tchobanoglous and Schroeder 1987; Harriott 2002). By using a range of ideal reactor models, we will be able to approximate the behavior of a range of lentic—slow moving—waterbodies including lakes and ponds and faster moving lotic systems such as rivers and streams.

9.1.1 Batch Reactor

The batch reactor is conceptually the simplest model we'll consider since there are no net flows entering or exiting the system. In this sense, a batch reactor is closed with respect to the transfer of matter with the surrounding environment. In fact, we've already worked with this type of system in our hypothetical experiment used to collect data to determine the order of various reactions in Chapter 5. In a batch reactor, the volume is filled, reactions occur in the reactor, and then the volume is discharged. Once processing begins, flow doesn't enter or leave the reactor, so the volume is fixed. Likewise, the contents of a batch reactor are considered completely mixed, so we can assume the contents are homogenous. The general mass balance for a batch reactor of volume, V, with a reaction, r, is given in Equation 9.1.

$$\text{Equation 9.1} \quad Q_{in}C_{in} - Q_{out}C_{out} \pm rV = V\frac{dC}{dt}$$

Since there's no flow of water into or out of V, $Q_{in} = Q_{out} = 0$; therefore, we can simplify this equation (Equation 9.2). We'll consider a negative reaction rate term to indicate the destruction or removal of a constituent, so we've dropped the "+/−" from the reaction term.

$$\text{Equation 9.2} \quad r = \frac{dC}{dt}$$

The expression in Equation 9.2 might look familiar since we've actually used it previously to describe the way the concentration of a solute changes over time in a

DOI: 10.1201/9781003289630-10

batch reactor for zero- and first-order reactions (Tchobanoglous and Schroeder 1987; Harriott 2002).

9.1.2 CONTINUOUS FLOW STIRRED TANK REACTOR (CFSTR)

A sketch of a CFSTR is presented in Figure 9.1. In a CFSTR, constituents (chemicals, sediment, *etc.*) flow into the reactor at a concentration, C_{in}, where they remain for a period of time equal to the hydraulic residence time. The contents of the reactor are assumed to be completely mixed. As a result, the concentrations of reactants in a CFSTR are the same throughout the entire reactor volume. While in the CFSTR, constituents can react prior to being discharged from the reactor. Since the contents are completely mixed, the concentration in the discharge from a CFSTR is the same as the concentration inside the reactor. Any fluid element has an equal chance of being discharged from the CFSTR at any time, so we consider the average time a fluid element remains in the reactor.

To simplify our analysis of this system, we can assume that Q_{in} and Q_{out} are equal. Finally, it's reasonable to assume that the volume of the reactor is fixed. Consequently, the rate of accumulation can be written as the product of the reactor volume and the change in concentration over time, dC/dt. A special case for the accumulation term is when there is no accumulation in the control volume. This is known as steady-state condition and dC/dt = 0. The mass balance expression for a CFSTR operating under steady-state hydraulic conditions with a fixed vessel volume is given in Equation 9.3.

$$\text{Equation 9.3} \quad Q_{in}C_{in} - Q_{out}C_{out} \pm rV = V\frac{dC}{dt}$$

One way to study the way water moves in a reactor is to use a tracer—a conservative substance (r = 0) that can be tracked as it moves through the system. Dyes are often used as tracers while radioisotopes (^{82}Br, ^{131}I, and ^{32}P) are used to study flow in more complicated systems. In a tracer study of a CFSTR, tracer is added continuously to the reactor at the inlet with a concentration, C_{in}, when t = 0. The reactor effluent is

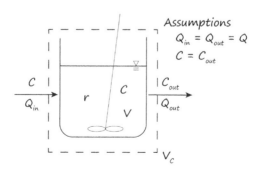

FIGURE 9.1   Schematic diagram of a CFSTR.

then monitored for the presence of the tracer. The mass balance expression for this case is given in Equation 9.4.

$$\text{Equation 9.4} \quad QC_{in} - QC = V\frac{dC}{dt}$$

From Equation 9.4, we can develop an expression for the concentration, C, over time. Since the flow rate, Q, is constant, we'll begin by pulling the flow out of both terms on the left-hand side (Equation 9.5).

$$\text{Equation 9.5} \quad Q(C_{in} - C) = V\frac{dC}{dt}$$

Remember that we also assumed that the volume, V, is constant, so we can gather it with the other constant, Q (Equation 9.6).

$$\text{Equation 9.6} \quad \frac{Q}{V}(C_{in} - C) = \frac{dC}{dt}$$

To begin our solution, we need to separate the differential terms—dt and dC—on opposite sides (Equation 9.7).

$$\text{Equation 9.7} \quad \frac{Q}{V}dt = \frac{dC}{(C_{in} - C)}$$

This equation can't be solved using algebra. Rather, it needs to be integrated (Equation 9.8).

$$\text{Equation 9.8} \quad \frac{Q}{V}\int_0^t dt = \int_{C_{in}}^C \frac{dC}{(C_{in} - C)}$$

To arrive at a closed-form solution, we need to specify a set of boundary conditions. In this case, our boundary conditions include $C = C_{in}$ when $t = 0$, and at some time, t, the concentration is C.

To simplify this analysis a little more, let's define the hydraulic retention time, θ_h, as in Equation 9.9.

$$\text{Equation 9.9} \quad \theta_h = \frac{V}{Q}$$

Using dimensional analysis, θ_h has units of time and is constant for a fixed volume and flowrate. After substituting Equation 9.8 into Equation 9.9, we can write Equation 9.10.

$$\text{Equation 9.10} \quad \frac{1}{\theta_h}\int_0^t dt = \int_0^C \frac{dC}{(C_{in} - C)}$$

The solution to the right-hand side of Equation 9.10 is actually of the general form that was used to find the concentration over time for a first-order reaction in Chapter 5. By integrating both sides of Equation 9.10, we can write a solution (Equation 9.11) (Chau 2018).

$$\text{Equation 9.11} \quad \frac{t}{\theta_h} = -\left(\ln(C_{in} - C) - \ln(C_{in})\right)$$

We can simplify Equation 9.11 further by using the quotient property of logs (Equation 9.12).

$$\text{Equation 9.12} \quad -\frac{t}{\theta_h} = \ln\left(\frac{C_{in} - C}{C_{in}}\right)$$

The normalized concentration over time (Equation 9.13) is finally found by taking the exponential of both sides of Equation 9.12 and preforming a little algebra (Equation 9.13 through Equation 9.16).

$$\text{Equation 9.13} \quad e^{-\left(\frac{t}{\theta_h}\right)} = \frac{C_{in} - C}{C_{in}}$$

$$\text{Equation 9.14} \quad C_{in} e^{-\left(\frac{t}{\theta_h}\right)} = C_{in} - C$$

$$\text{Equation 9.15} \quad C = C_{in} - C_{in} e^{-\left(\frac{t}{\theta_h}\right)} = C_{in}\left(1 - e^{-\left(\frac{t}{\theta_h}\right)}\right)$$

$$\text{Equation 9.16} \quad \frac{C}{C_{in}} = 1 - e^{-\left(\frac{t}{\theta_h}\right)}$$

Now, let's take a look at a plot of C/C_{in} versus t/θ_h (Figure 9.2). For a CFSTR, steady-state hydraulic conditions are established after exchanging about three reactor volumes (Tchobanoglous and Schroeder 1987; Harriott 2002).

9.1.3 First-Order Removal in a CFSTR

Now that we understand how water moves in a CFSTR, we can consider the effect(s) of a first-order reaction on the performance of a CFSTR. The mass balance

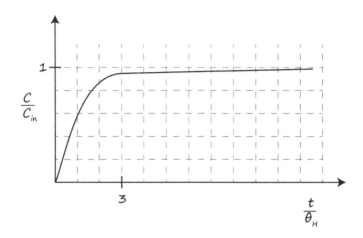

FIGURE 9.2 Plot of normalized concentration and time for a tracer in a CFSTR.

expression for a CFSTR operating under steady-state hydraulic conditions with first-order removal (r = –kC) is given in Equation 9.17.

$$\text{Equation 9.17} \quad QC_{in} - QC - kCV = V\frac{dC}{dt}$$

The intermediate steps to find the concentration over time are presented in Equation 9.18 through Equation 9.20.

$$\text{Equation 9.18} \quad \frac{Q}{V}C_{in} - \frac{Q}{V}C - kC = \frac{dC}{dt}$$

$$\text{Equation 9.19} \quad \frac{C_{in}}{\theta_h} - C\left(\frac{1}{\theta_h} + k\right) = \frac{dC}{dt}$$

$$\text{Equation 9.20} \quad \frac{dC}{dt} + C\left(\frac{1}{\theta_h} + k\right) - \frac{C_{in}}{\theta_h} = 0$$

The solution to this homogenous first-order ordinary differential equation is given in Equation 9.21. In the context of this work, we're more concerned with the physical significance of the outcome and less with the mechanics of solving differential equations.

$$\text{Equation 9.21} \quad C = \frac{C_{in}}{1 + k\theta_h}\left(1 - e^{-\frac{(1+k\theta_h)t}{\theta_h}}\right)$$

The expression in Equation 9.21 appears to be more complicated than those we've worked with in the past; however, we can develop a plot of normalized concentration versus normalized time (Figure 9.3) by considering the behavior of Equation 9.21 at its extremes. Remember that we set boundary conditions for this system earlier. Initially, the concentration of C in the reactor effluent is zero. As time goes to

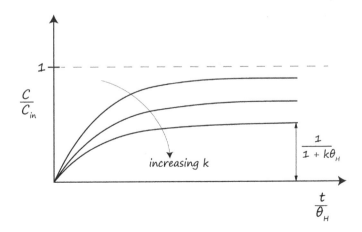

FIGURE 9.3 Plot of normalized concentration and time for first-order removal in a CFSTR.

infinity, the steady-state concentration in a CFSTR with first-order removal is given in Equation 9.22. The exponential of a negative goes to zero as time goes to infinity and we're left with the constant term in Equation 9.22.

$$\text{Equation 9.22} \quad \lim_{t \to \infty} C = \lim_{t \to \infty}\left(\frac{C_{in}}{1+k\theta_h}\left(1 - e^{-\frac{(1+k\theta_h)t}{\theta_h}}\right)\right) = \frac{C_{in}}{1+k\theta_h}$$

We can also see how the steady-state effluent concentration in a CFSTR is impacted by a change in the reaction rate constant, k. In this system, a higher reaction rate constant will yield a lower steady-state concentration in the reactor discharge (Tchobanoglous and Schroeder 1987; Harriott 2002).

The steady-state concentration in the CFSTR can also be found directly from the mass balance equation by assuming that the concentration doesn't change over time, as in Equation 9.23.

$$\text{Equation 9.23} \quad QC_{in} - QC - kCV = 0$$

If we gather like terms and solve for C/C_{in}, the solution to Equation 9.23 is exactly what we determined earlier in Equation 9.22.

CFSTR models are often used to describe the movement of mass in ponds and lakes. This approach has a number of significant limitations including the need to assume complete mixing. Clearly, in natural systems, we don't have the ability to ensure that the entire lake volume is completely mixed and the concentration of constituents is homogenously distributed throughout the volume.

9.1.4 CFSTRs IN SERIES

There are times when we might want to describe the way concentration changes as water flows through a series of environmental compartments. For example, we might want to describe the movement of water in a river reach as a series of CFSTRs, as shown schematically in Figure 9.4. These could be a series of step pools in a stream. Since we're applying the CFSTR model to this system, we assume that there is complete mixing in each reactor. We can write the steady-state effluent concentration for first-order removal from the first CFSTR (Equation 9.24).

$$\text{Equation 9.24} \quad C_1 = \frac{C_{in}}{1+k_1\theta_{h_1}}$$

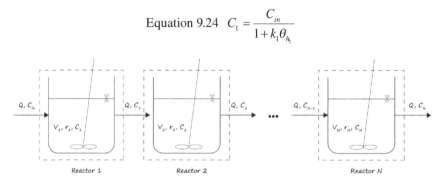

FIGURE 9.4   A schematic of multiple CFSTRs in series.

The mass balance equation for the steady-state concentration in the second CFSTR is given in Equation 9.25.

$$\text{Equation 9.25} \quad Q_2 C_2 - Q_1 C_1 - k_2 C_2 V_2 = 0$$

The solution to Equation 9.25 has a form that we saw previously (Equation 9.26).

$$\text{Equation 9.26} \quad C_2 = \frac{C_1}{\left(1 + k_2 \theta_{h_2}\right)}$$

Since we know C_1 (Equation 9.24), we can include C_{in} and the dimensions and reaction rate in Reactor 1 in our solution for C_2.

$$\text{Equation 9.27} \quad C_2 = \frac{C_{in}}{\left(1 + k_1 \theta_{h_1}\right)\left(1 + k_2 \theta_{h_2}\right)}$$

Based on Equation 9.27, we can write a general solution for the steady-state concentration from N CFSTRs in series (Equation 9.28) (Tchobanoglous and Schroeder 1987).

$$\text{Equation 9.28} \quad C_N = \frac{C_{in}}{\prod_{i=1}^{N}\left(1 + k_i \theta_{h_i}\right)}$$

9.1.5 Plug Flow Reactor (PFR)

A differential element in an ideal PFR is illustrated in Figure 9.5. Unlike batch reactors and CFSTRs, ideal PFRs feature compete lateral mixing (y–z plane) while the reactor is unmixed in the longitudinal direction (along the x-axis in Figure 9.5). This

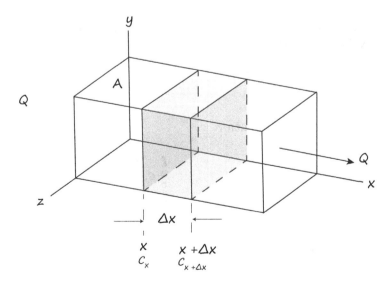

FIGURE 9.5 A differential element of a plug flow reactor.

system can be thought of as a series of discrete beakers or boxes on a conveyor belt. In comparison to CFSTRs, every fluid element in a PFR resides in the reactor for an equal period of time. As we've done when considering other reactor models, we can assume the flow rates in and out of the reactor are constant.

9.1.6 Tracer Experiment (r = 0) in a PFR

To establish a better understanding of the way water moves through a PFR, we can perform a tracer study in a manner similar to the approach we used for the CFSTR. We'll begin by pumping clean water into the reactor until the reactor is full and the flow rate is constant. Then, we'll add a tracer to the reactor inlet at a concentration C_T for a period of time equal to $1\theta_h$. A plot of the influent tracer concentration is presented in Figure 9.6 (left). After filling the reactor with tracer, we'll flush the reactor with clean water while we observe the concentration in the effluent (Figure 9.6, right). In an ideal PFR, the tracer moves as a slug with complete mixing for a period of one hydraulic detention time. After $1\theta_h$, the tracer concentration in the PFR's discharge returns to zero.

9.1.7 First-Order Removal in a PFR

We've seen that constituents move through a PFR as a slug; that is, there is complete mixing in a plane perpendicular to the direction of flow. Now, we need to develop an idea of how concentration changes along the length of a PFR (in the x-direction). Let's start by looking at a differential element of thickness, Δx, along the length of an ideal PFR, as shown previously in Figure 9.5. The mass balance equation for this case is given in Equation 9.29.

$$\text{Equation 9.29} \quad QC_x - QC_{x+\Delta x} \pm r\Delta V = \Delta V \frac{\Delta C}{\Delta t}$$

For a reactor with a constant cross-sectional area, A, the volume is the product of A and Δx (Equation 9.30).

$$\text{Equation 9.30} \quad \Delta V = A\Delta x$$

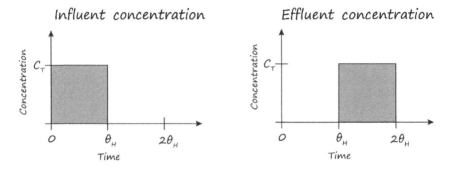

FIGURE 9.6   Influent and effluent concentrations for a tracer in an ideal PFR.

A bit of algebra is needed to find a form of this equation that we can solve for the concentration, C, as in Equation 9.31 through Equation 9.32. We can divide both sides by ΔV.

$$\text{Equation 9.31} \quad \frac{Q}{A}\left(\frac{C_x - C_{x+\Delta x}}{\Delta x}\right) \pm r = \frac{\Delta C}{\Delta t}$$

$$\text{Equation 9.32} \quad -\frac{Q}{A}\left(\frac{C_{x+\Delta x} - C_x}{\Delta x}\right) \pm r = \frac{\Delta C}{\Delta t}$$

$$\text{Equation 9.33} \quad -\frac{Q}{A}\left(\frac{\Delta C}{\Delta x}\right) \pm r = \frac{\Delta C}{\Delta t}$$

We can define a small change in the hydraulic retention time according to Equation 9.34.

$$\text{Equation 9.34} \quad \Delta\theta_h = \frac{\Delta V}{Q} = \frac{A\Delta x}{Q}$$

We can then rewrite the left-hand side of Equation 9.35 as a small change in θ_h, $\Delta\theta_h$.

$$\text{Equation 9.35} \quad -\frac{\Delta C}{\Delta\theta_h} \pm r = \frac{\Delta C}{\Delta t}$$

For a steady-state concentration, $\Delta C/\Delta t = 0$. This allows us to write the reaction rate for a PFR operating at a steady-state concentration according to Equation 9.36.

$$\text{Equation 9.36} \quad r = \frac{\Delta C}{\Delta\theta_h}$$

Equation 9.36 should look familiar. In fact, this is very close to the result we developed for first-order removal in a batch reactor. We can adapt this solution to fit our analysis of first-order removal in an ideal PFR, where we'll replace "t" with "θ_h" (Equation 9.37).

$$\text{Equation 9.37} \quad \frac{C}{C_{in}} = e^{-k\theta_h} = e^{-kA\frac{\Delta x}{Q}}$$

For an ideal PFR with a length L, the effluent concentration over distance is presented in Figure 9.7. To achieve a lower outlet concentration in an ideal PFR with first-order removal, a longer reactor should be used. In comparison, a greater reduction in the initial concentration was obtained in a batch reactor with first-order removal when more time was allowed for the reaction to proceed (Tchobanoglous and Schroeder 1987; Harriott 2002).

Example: CAFO Blowout Part 1

Confined animal feeding operations (CAFOs) are known to be sources of biochemical oxygen demand that can have adverse impacts on downstream water quality. Consider a swine CAFO with a BOD of 10,000 mg/L (Burkholder *et al.*

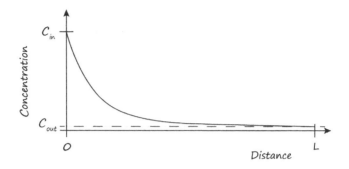

FIGURE 9.7   Concentration versus distance for first-order removal in an ideal PFR.

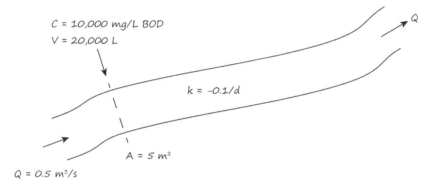

FIGURE 9.8   Swine CAFO accidental discharge into a regional waterway.

2007). Imagine that 20,000 L is accidently discharged as a slug into a regional waterway that flows at 0.5 m/s. Assume the channel is fairly uniform with a typical cross-sectional area of 5 m². A sketch of this system is presented in Figure 9.8. If the first-order reaction rate constant for BOD removal is 1.2/d (Webb and Archer 1997), how far downstream will the slug of BOD have to travel for the BOD in the river reach to be 10% of its initial value? How far will it have to travel to be 1% of the initial concentration?

Let's assume we have steady-state flow and there are no other sources of BOD entering the river reach. Let's also consider that the CAFO waste was discharged into the stream as a discrete slug. In relationship to the dimensions of the channel, the 20,000 L (20 m³) waste slug would occupy a 4 m long stream segment. In this case, we want to track the BOD concentration of the CAFO waste slug as it moves downstream. The control volume would be the waste slug (5 m² x 4 m long).

We know the relationship between the concentration, flowrate, and channel dimensions for first-order removal in a PFR (Equation 9.38). To find the distance needed for the BOD concentration to be reduced to 10% of its initial value, we can solve for Δx.

$$\text{Equation 9.38} \quad \Delta x = -\left(\frac{Q}{Ak}\right)\ln\left(\left(\frac{C}{C_{in}}\right)\right) = -\left(\frac{0.5\frac{m^3}{s}}{(5m^2)\left(\frac{1.2}{d}\right)\left(\frac{1d}{86,400s}\right)}\right)\ln(0.1)$$

$$= 16,579m \cong 16.6km \cong 10.3mi$$

We can find the distance for the BOD concentration to degrade to 1% of its initial value using the same approach. The slug would have to travel 33,157 km (~20.6 mi) for 99% of the BOD to be removed.

9.2 COMPARING CFSTR AND PFR PERFORMANCE

We need a basis to compare the performance of CFSTRs and PFRs. In this case, we could compare the two reactors based on the volume of each system needed to reach a particular effluent concentration. However, since these are flowing water systems, a comparison based on hydraulic retention time is more rigorous. For this analysis, we'll also assume that steady-state hydraulic and concentration conditions are applicable and that a first-order removal reaction occurs in both reactors. We can do this by solving Equation 9.22 for θ_h in a CFSTR. Likewise, for a PFR operating under steady-state conditions with first-order removal, we can solve Equation 9.36 for θ_h. To complete the comparison of these reactor systems, let's target a removal of 90% and assume k = 0.1/min. After a little number crunching, we'll find that θ_h for the CFSTR is 90 min (Equation 9.39) and θ_h for the PFR is 23 min (Equation 9.40).

$$\text{Equation 9.39} \quad \theta_h = \left(\frac{1}{k}\right)\left(\frac{C_{in}}{C}-1\right) = \left(\frac{min}{0.1}\right)(10-1) = 90\,min$$

$$\text{Equation 9.40} \quad \theta_h = -\left(\frac{1}{k}\right)\ln\left(\frac{C}{C_{in}}\right) = -\left(\frac{min}{0.1}\right)\ln(0.1) = 23\,min$$

What does this mean in practical terms? For a CFSTR and a PFR with equal flow rate, reaction order, reaction rate constant, and removal efficiency, less reactor volume is needed when using plug flow hydraulics (Davis and Davis 2003; Tchobanoglous and Schroeder 1987).

Example: CAFO Blowout Part 2

Consider a case where the CAFO discharge occurs 8 km upstream from a lake with a volume of 10^5 m³ (Figure 9.9). Find the BOD concentration in the stream that drains the lake. In this case, we'll perform a mass balance on BOD in the lake. Let's assume steady-state flow with no additional sources of BOD in the streams or the lake. Likewise, we need to assume complete mixing in the lake.

We know how to describe the steady-state concentration in the outlet of a CFSTR with first-order removal (Equation 9.22). However, we need to determine the BOD concentration that enters the lake after moving 8 km from its source. To do this, we'll treat the 8 km long river reach as a plug flow reactor.

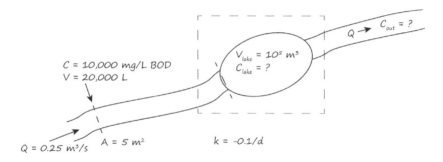

FIGURE 9.9   Swine CAFO discharge into a regional waterway that feeds a lake.

We can use the equation we developed earlier for first-order removal in a PFR under steady-state flow conditions (Equation 9.41), where Δx = 8 km.

$$\text{Equation 9.41}\quad C = C_{in}e^{-k\theta_h} = \left(10,000\frac{mg}{L}\right)e^{-\left(\frac{1.2}{d}\right)\left(\frac{1d}{86,400s}\right)\left(5m^2\right)\frac{\left(8\cdot10^3\,m\right)}{5\frac{m^3}{s}}}$$

$$= 8,948\frac{mg}{L}$$

To find the BOD concentration in the lake, we need to determine the hydraulic retention time for the lake (Equation 9.42).

$$\text{Equation 9.42}\quad \theta_h = \frac{10^5 m^3}{5\frac{m^3}{s}} = 2.0\cdot10^4 s \cong 0.23d$$

Now, we can find the BOD concentration in the lake and the stream that drains the lake (Equation 9.43), since C in an ideal CFSTR is equal to C_{out}.

$$\text{Equation 9.43}\quad C = \frac{C_{in}}{1+k\theta_h} = \frac{8,948\frac{mg}{L}}{1+\left(\frac{1.2}{d}\right)(0.23d)} = 7,013\frac{mg}{L} \cong 7,000\frac{mg}{L}$$

Example: CAFO Blowout Part 3

If the streams that feed and drain the lake have roughly the same dimensions, at what distance downstream from the lake will the BOD concentration be reduced to 1,000 mg/L (10% of the initial spill concentration)?

In this case, the BOD concentration flowing into the downstream river reach is ~7,000 mg/L and our target concentration is 1,000 mg/L. We can use the same approach we used in CAFO Blowout Part 1 (Equation 9.44).

$$\text{Equation 9.44} \quad \Delta x = -\left(\frac{0.5\frac{m^3}{s}}{\left(5m^2\right)\left(\frac{1.2}{d}\right)\left(\frac{1d}{86,400s}\right)}\right)\ln\left(\frac{1,000}{7,000}\right)$$

$$= 14,011m \cong 14km \cong 8.7mi$$

Recall that it took 10.3 km of river reach to achieve a 90% reduction in CAFO Blowout Part 1. When the CFSTR was added, treatment occurred over a time period that was proportional to the hydraulic residence time. Ultimately, the CFSTR resulted in a 16% reduction in the length of the stream needed to reach 90% removal.

REFERENCES

Chau, K. (2018). *Differential Equations in Engineering and Mechanics, Theory and Applications*, CRC Press, Boca Raton, FL, 1790 pp.

Davis, M., and R. Davis (2003). *Fundamentals of Chemical Reaction Engineering. McGraw-Hill Chemical Engineering Series*, McGraw-Hill Higher Education, New York, NY, 384 pp.

Harriott, P. (2002). *Chemical Reactor Design*, CRC Press, Boca Raton, FL, 448 pp.

Tchobanoglous, G., and E. Schroeder (1987). *Water Quality*, Addison-Wesley, Reading, MA, 768 pp.

10 Departure from Ideal Conditions

10.1 PRACTICAL DIFFERENCES BETWEEN IDEAL REACTORS AND NATURAL SYSTEMS

Up to this point, we've focused primarily on the behavior of reactive and conservative substances while spending relatively little time examining the effects of flow on mixing, hydraulic residence time, *etc.* Now, we need to examine the ways in which non-ideal flow patterns affect these parameters and—as a consequence—our ability to predict the behavior of transport and reaction mechanisms in natural systems. To better inform our modeling of natural systems, it's important to understand the ways in which particular conditions can affect the distribution of flow in a reactor system.

When we modeled a lake as a CFSTR, we assumed that the reactor volume was well mixed and the concentration was uniform throughout. Similarly, we assumed complete lateral mixing for an ideal PFR. In natural systems, our ability to ensure complete mixing is complicated by factors that include thermal stratification, often low inlet and outlet flowrates relative to lake volume, and an incomplete understanding of water flow in the larger system. Other factors, including variable temperatures, variations in wind-induced mixing, the influence of aquatic vegetation and woody debris on water flow, and other nature-controlled operating parameters, cause less than ideal flow in natural systems. For example, the log in Figure 8.1 (top) collected sediment and also altered the flow of water in the channel.

Initially, we considered channel cross sections to be uniform and symmetrical about a central axis. We also assumed that the channel cross section remains approximately the same along the length of a channel. In reality, we know that these two conditions are rarely seen in natural systems. Consequently, when making calculations for parameters related to flow, turbulence, *etc.*, the compound cross section approach is necessary.

We've seen that the movement of sediment is a key feature of alluvial channels. When sediment accumulates in a channel, the capacity of the waterway is reduced. This results in an effective hydraulic retention time that's less than that predicted based on channel dimensions and flowrate.

Example: Determining the Impact of Sediment on Hydraulic Retention Time in a First-Order Stream

Consider a cross section of a first-order stream in Knox County, IL (USA), that's presented in Figure 10.1 (top). The area of this cross section, 1.180 m², was found by treating this as a compound cross section (Figure 10.1, bottom). Like many low-order streams in the Midwestern US, Forman Creek contains a substantial

 DOI: 10.1201/9781003289630-11

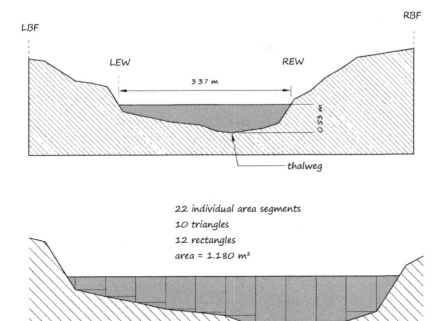

FIGURE 10.1   A cross section of Forman Creek, a first-order channel in Knox County, IL, USA (top). The area is found by treating this as a compound cross section (bottom).

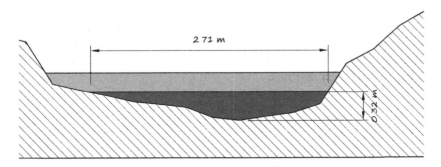

FIGURE 10.2   A cross section of Forman Creek with sediment layer. The presence of the sediment layer decreases the effective hydraulic retention time of the channel.

sediment layer (Figure 10.2). By using the compound cross section approach, the cross-sectional area of the sediment layer was 0.507 m². In this case, sediment occupies nearly 43% of the stream cross section.

To see how the presence of the sediment layer impacts retention time in the channel, we need to develop expressions for θ_h in the absence of sediment (a "clean" channel) and for θ_h when the sediment layer is present. Let's treat this case as a PFR and assume the channel has constant conveyance (flowrate). We'll also compare retention times based on an equal distance along the stream, Δx. We can write the hydraulic retention time for the clean channel as shown in Equation 10.1.

$$\text{Equation 10.1} \quad \theta_{h-clean} = \frac{A_{clean}\Delta x}{Q} = (1.180m)\frac{\Delta x}{Q}$$

where A_{clean} is the cross-sectional area of the clean channel.

The hydraulic retention time for the sediment-laden channel can be written in a similar fashion (Equation 10.2).

$$\text{Equation 10.2} \quad \theta_{h-filled} = \frac{A_{filled}\Delta x}{Q} = (1.180m - 0.507m)\frac{\Delta x}{Q} = (0.673)\frac{\Delta x}{Q}$$

By calculating the ratio of retention times for the clean and filled channels (Equation 10.3), we can find that the presence of sediment in the channel reduced the hydraulic retention time by 57%.

$$\text{Equation 10.3} \quad \frac{\theta_{h-filled}}{\theta_{h-clean}} = \frac{0.673}{1.180} = 0.570$$

10.2   NON-IDEAL FLOW

Now, let's consider a few particular causes of non-ideal flow that are seen in engineered systems as well as in streams and lakes. Hydraulic short circuiting— when water flows more or less directly from the inlet to the outlet of a reactor (Figure 10.3)—is a common cause of non-ideal flow. For example, the velocity and momentum of water from an inlet with a small cross-sectional area can be high enough to force water directly from the inlet to the outlet. This results in a hydraulic retention time that is significantly lower than the value calculated from basin dimensions and water flow rate.

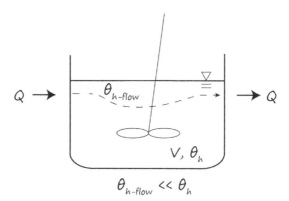

FIGURE 10.3   Hydraulic short circuiting can cause significant differences in predicted and actual behavior in ideal reactor models (adapted from Winterbottom and King 1999).

In ideal reactors, short circuiting can reduce the time needed for (bio)chemical reactions to reach desired endpoints and can decrease the time needed for effective solids removal through gravity settling. Dead zones—areas in a reactor that receive little to no turnover of water—can also develop. Other impacts of short circuiting include excessive settling of sediment and chemical precipitates—often in reactor dead zones. Additional adverse impacts to downstream waters can occur when solids and soluble constituents are insufficiently treated or removed. In other cases, channelized flow can develop in sediment, which also results in a deviation from calculated retention times.

One of the most obvious signs of short circuiting is an appreciable difference between predicted and actual reactor performance. In engineered systems, this is frequently obvious. However, in natural systems, it can be difficult to fully account for changes in water flow rates and channel dimensions. For example, Boutilier *et al.* (2011) reported that high flow rates and the presence of dense stands of cattails had an adverse impact on their ability to model the fate and transport of *E. coli* in treatment wetlands. They also noted the inherent challenges in modeling a dynamic system that was open to the environment due to the large number of potentially rate-limiting variables. Likewise, they found that flowrate changed as the botanical community matured.

Rose *et al.* (2004) showed the impact of inlet-to-outlet short circuiting on phosphorus removal in Pike Lake (Washington County, WI, USA). In this case, they noted a 65–77% annual difference between predicted and measured phosphorus discharges from Pike Lake (~187 ha with a mean depth of ~4.1 m). For example, in one year, they predicted a 181 kg (400 lb) discharge form the lake, while the actual load was 1,134 kg (2,500 lb).

Thermal stratification in a waterbody is another factor that is known to contribute to hydraulic short circuiting. For example, in winter, cold water will settle to the bottom of a waterbody. Warm water takes the path of least resistance and rides on top of the thin dividing layer between thermal strata, which results in short circuiting. If the stream contains elevated BOD, the biosolids settle to the bottom of the waterbody and can create an anoxic zone which has additional implications for the physicochemical and biological characteristics of the waterway. Likewise, temperature stratification can result in overall predicted reaction rates that are lower than anticipated.

Recall that in a CFSTR, every element of a fluid in the reactor volume has an equal chance of being discharged from the reactor at any time. Consequently, each fluid packet has its own hydraulic retention time. We accommodate this fact by using an average θ_h. Based on this approach, you might think a reasonable strategy would be to average the retention times of each distinctive flow path. This approach might work under certain limited circumstances; however, it fails to adequately incorporate the effects of the dead spaces that result in regions that receive very little bulk flow (Figure 10.4, top). In some cases, the effects of temperature stratification are mitigated in the upper layers of a waterbody by wind-driven circulation (Figure 10.4, bottom). However, this typically doesn't aid in mixing deeper layers. We also know that the reaction rate, r, is affected by temperature. Consequently, reactions in thermally stratified systems will take place at different rates in regions with different temperatures.

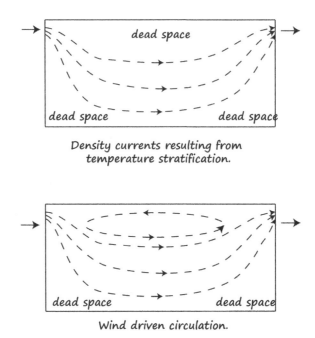

Density currents resulting from
temperature stratification.

Wind driven circulation.

FIGURE 10.4 Schematic flow stream lines for a lake that's fed and drained by a river with and without wind-driven circulation of the near-surface layer (top and bottom, respectively) (adapted from Winterbottom and King 1999).

The presence of a temperature gradient between layers can also require a way to account for thermally driven transport between strata. While the mathematics needed to describe this case are outside the scope of this book, the impacts of this movement of mass in water systems can be considered qualitatively. In the presence of a temperature gradient, energy will flow from regions of high temperature to areas of lower temperature. This can also drive the movement of soluble water constituents. In these systems, dispersion, the movement of mass from regions of high concentration to regions of lower concentration due to differences in flow velocities along various flow paths, can also become an important mechanism for the movement of constituents in water. Mixing between flow paths due to molecular diffusion can also contribute to mass transport via dispersion. In engineered systems, mixers are used to manage thermal stratification in treatment reactors.

To address the issue of hydraulic short circuiting, the inlet dimensions and influent water velocity in engineered or natural systems should be considered when examining the potential for non-ideal flow. This might include introducing water into a lake via a broad channel that contains large stones or boulders for energy dissipation.

The accumulation of sediment in alluvial channels also often results in channelization. A terrestrial analog for channelization is the formation of rills that are created by soil erosion. In this case, flow is directed through discrete channels that develop in loose alluvial deposits. These channels have dimensions that can be substantially different from the larger main channel. Channelization is often exacerbated by the presence of baffles in engineered systems or physical obstructions to flow (*e.g.*, woody debris) in natural systems.

10.3 TRACER STUDIES TO IDENTIFY NON-IDEAL FLOW

Earlier, we examined the response of an ideal PFR to the input of a tracer lasting for a time period of θ_h and observed that there was a $1\theta_h$ delay in the appearance of the tracer in the reactor effluent. Now, consider the addition of a tracer "spike" into an ideal PFR (Figure 10.5). In this system, the spike tracer input travels across the PFR and is observed as a spike in the reactor effluent at $t = \theta_h$. However, it's been shown that the same spike input into natural systems, including wetlands, can yield an observed hydraulic residence time, $\theta_{h\text{-obs}}$, that's less than θ_h. In systems that deviate significantly from ideal conditions, the PFR outlet concentration will be wider and lower in peak amplitude, as shown in Figure 10.5.

Consider a pond (Figure 10.6, top) that has two different flow paths. In this case, a fluid element that follows Path 1 will spend less time in the pond than one that follows Path 2. Path 1 exhibits behavior that's consistent with hydraulic short circuiting, while Path 2 is closer to ideal flow. Regardless, the presence of two different flow paths will necessarily complicate the calculation of a representative hydraulic retention time. Now consider the same pond with the addition of a zone of dense vegetation (Figure 10.6, middle) where the water velocity is close to zero. In practice, large differences between the observed and calculated retention times often indicate the presence of a dead zone (Persson 1999). A final case of non-ideal flow in the pond is shown in Figure 10.6 (bottom), where the presence of woody debris further complicates our determination of an appropriate hydraulic residence time. In this instance, the direction of flow along Path 2 is actually redirected toward the inlet, further increasing the actual amount of time a fluid element remains in the system.

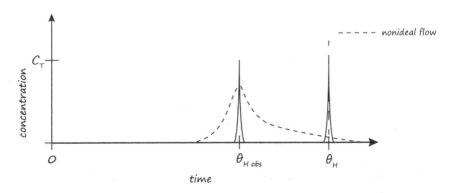

FIGURE 10.5   Ideal and non-ideal PRF responses to a spike tracer input (Winterbottom and King 1999).

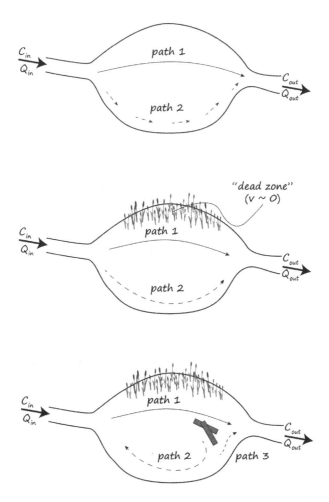

FIGURE 10.6   Different flow paths in a pond (top) and a wetland that contains dense vegetation and woody debris (bottom).

10.4   RESIDENCE TIME DISTRIBUTION

One approach that can be used to account for differences in the anticipated/calculated hydraulic retention time and behavior we observe in tracer studies is the residence time distribution (RTD). The RTD is the probability that water will flow along a particular path. For a constant water flowrate, we can use the RTD to weight the overall average effluent concentration, $C_{weighted}$. For n discrete flow paths, we need to know the RTD for each discrete flow path so we can calculate a corresponding hydraulic retention time for each path. If we treat this system as a PFR, $C_{weighted}$ can be calculated according to Equation 10.4.

Equation 10.4 $C_{weighted} = RTD_1 C_{in} e^{-k\theta_{h_1}} + RTD_2 C_{in} e^{-k\theta_{h_2}} + \ldots + RTD_{n-1} C_{in} e^{-k\theta_{h_{n-1}}}$
$$+ RTD_n C_{in} e^{-k\theta_{h_n}}$$

The RTD can also be a continuous function. In this case, Equation 10.4 must be written in differential form (Equation 10.5).

$$\text{Equation 10.5} \quad C_{weighted} = C_{in} \int RTD(t)e^{-k\theta}\,dt$$

In most cases, residence time distributions used to model the continuous case are derived through some form of modeling.

Example: Applying the RTD Approach to a PFR with Two Flow Paths

Consider a pond with plug flow hydraulics where water can take one of three paths. A first-order reaction with a rate constant of 0.01/d takes place in the reactor. Path 1 has a hydraulic retention time of 8 d, while fluid elements that follow Path 2 and Path 3 remain in the system for 6 hr (0.25 d) and 1 d, respectively. If there's a 20% chance that water will take Path 1, a 30% chance it will follow Path 2, and a 50% chance it will take Path 3, what fraction of the influent concentration is removed?

We can build an expression for the normalized weighted concentration by applying Equation 10.6.

$$\text{Equation 10.6} \quad \frac{C}{C_{in}} = (0.2)e^{-\left(\frac{0.1}{d}\right)(8d)} + (0.3)e^{-\left(\frac{0.1}{d}\right)(0.25d)} + (0.5)e^{-\left(\frac{0.1}{d}\right)(1d)} = 0.84$$

What if we just averaged the retention times for the three paths (Equation 10.7) and calculated the removal (Equation 10.8)?

$$\text{Equation 10.7} \quad \theta_h = \frac{8d + 0.25d + 1d}{3} = 3.083d \cong 3.1d$$

$$\text{Equation 10.8} \quad \frac{C}{C_{in}} = e^{-\left(\frac{0.1}{d}\right)(3.1d)} = 0.73$$

Our removal is 13% better when we used the average retention time. From a practical point of view, using the average retention time resulted in an over-prediction of the removal in the system.

In the preceding case, we looked at a system that was best characterized using plug flow hydraulics. Now, let's consider the case where our system is best described as an ideal CFSTR. We already know a general solution for the first-order removal in a steady-state CFSTR. We can use this as the basis to write the concentration for a CFSTR with N flow paths that each has its own hydraulic retention time (Equation 10.9).

$$\text{Equation 10.9} \quad C_{weighted} = \frac{RTD_1 C_{in}}{1 + k\theta_{h_1}} + \frac{RTD_2 C_{in}}{1 + k\theta_{h_2}} + \ldots + \frac{RTD_{n-1} C_{in}}{1 + k\theta_{h_{n-1}}} + \frac{RTD_n C_{in}}{1 + k\theta_{h_n}}$$

REFERENCES

Boutilier, L., Jamieson, R., Gordon, R., and C. Lake (2011). "Modeling E. Coli Fate and Transport in Treatment Wetlands Using the Water Quality Analysis and Simulation Program," *Journal of Environmental Science and Health, Part A*, 46 (7), 680–691.

Burkholder, J., Libra, B., Weyer, P., Heathcote, S., Kolpin, D., Thorne, P., and M. Wichman (2007). "Impacts of Waste from Concentrated Animal Feeding Operations on Water Quality," *Environmental Health Perspectives*, 115 (2), 308–312.

Fogler, H. Scott (2005). *Elements of Chemical Reaction Engineering*, 4/e, Prentice Hall, Upper Saddle River, NJ, 1080 pp.

Holland, J., Martin, J., Granata, T., Bouchard, V., Quigley, M., and L. Brown (2004). "Effects of Wetland Depth and Flow Rate on Residence Time Distribution Characteristics," *Ecological Engineering*, 23 (3), 189–203, https://doi.org/10.1016/j.ecoleng.2004.09.003.

Persson, J., Somes, N., and T. Wong (1999). "Hydraulics Efficiency of Constructed Wetlands and Ponds," *Water Science and Technology*, 40 (3), 291–300.

Rodrigues, A. (2021). "Residence Time Distribution (RTD) Revisited," *Chemical Engineering Science*, 230, 116188, https://doi.org/10.1016/j.ces.2020.116188.

Rose, W., Robertson, D., and E. Mergener (2004). *Water Quality, Hydrology, and the Effects of Changes in Phosphorus Loading to Pike Lake, Washington County, Wisconsin, with Special Emphasis on Inlet-to-Outlet Short-Circuiting*, US Geological Survey Scientific Investigations Report 2004–5141, https://pubs.usgs.gov/sir/2004/5141/pdf/SIR_2004-5141.pdf.

Webb, J., and J. Archer (1994). "Pollution of Soils and Watercourses by Wastes from Livestock Production Systems," in Dewi, I., Axford, R., Marai, I., and H. Omed (Eds.), *Pollution in Livestock Production Systems*, CABI Publishing, Oxfordshire, UK, 189–204.

Werner, T., and R. Kadlec (2000). "Wetland Residence Time Distribution Modeling," *Ecological Engineering*, 15 (1–2), 77–90, https://doi.org/10.1016/S0925-8574(99)00036-1.

Winterbottom, J., and M. King (Eds.) (1999). *Reactor Design for Chemical Engineers*, 1/e, CRC Press, Boca Raton, FL, 454 pp.

11 Conversion Factors, Constants, and Reference Materials

Distance	1 m = 3.281 ft
Area	1 m^2 = 10.76 ft^2
Volume	1 m^3 = 35.32 ft^3 = 7.481 gal
Speed	1 m/s = 3.281 ft/s 1 km/hr = 0.6214 mi/hr
Flow rate	1 m^3/min = 35.32 ft^3/min = 264.2 gal/min
Pressure	1 kPa = 9.869 x 10^{-3} atm = 7.500 mm Hg
Mass	1 g = 2.205 x 10^{-3} lb mass
Temperature	T(°C) = 5/9[T(°F) − 32]
	T(K) = T(°C) + 273.15
Energy	1 J = 0.239 cal = 9.478 x 10^{-4} BTU
Absolute viscosity	1 poise = 1 dyne·s/cm^2 = 1 g/cm·s = 1/10
	Pa·s = 1/10 N·s/m^2
Kinematic viscosity	1 St = 10^{-4} m^2/s = 1 cm^2/s
Ideal gas constant	R = 1.987 cal/mole·K = 8.314 J/mole·K
Barometric pressure at sea level	101.325 kPa = 760.00 mm Hg = 1 atm
Activation energy for microbial systems	33,500 < E$_A$ < 50,000 J/mol
Acceleration due to gravity	g = 9.810 m/s^2 = 32.17 ft/s^2

DOI: 10.1201/9781003289630-12

PERIODIC TABLE OF THE ELEMENTS

Index

A

abiotic factors, 4, 103
absolute viscosity, 78, 80, 92, 139, *see also* viscosity
acid-base reactions, 49, 50, 53, 59
acid equilibrium constant, K_A, 49–50, *see also* equilibrium reaction constant
acidity, 31
active concentration, 29–31
activity coefficient, 30–31
advection, 19, 85–86, 93–95
alkalinity, 14, 31–34, 61
apparent color, 39–40, *see also* color

B

batch reactor, 43, 117–118, 125, *see also* reactors
bedload, 101–102, 107, 113
biochemical oxygen demand (BOD), 34–37, 125–129, 133
biological drift, 105
biotic components, 3–4, 97, 103–104
BOD_5, 35, 37, *see also* biochemical oxygen demand
BOD_u, 35, *see also* biochemical oxygen demand
Bronsted-Lowry acid, 49
buoyant force, 90–91

C

carbonate hardness, 33–34, *see also* hardness
carbonic acid, 50–55
CFSTRs, 117–123, 127–130, 133, 137, *see also* Continuous Flow Stirred Tank Reactor
CFSTRs in series, 122–123, *see also* reactors
characteristic length, 77, 81
chemical oxygen demand (COD), 34, 36–37
Clean Water Act, 16
closed system, 51, 54, *see also* thermodynamic systems
color, 38–40
common ion, 57–58
complexation, 58–61, *see also* coordinated compound
compound cross section, 106–107, 130–131
Confined Animal Feeding Operation (CAFO), 125–129
conservative substance, 118, *see also* tracer
constancy of flow, 97–98
Continuous Flow Stirred Tank Reactor (CFSTR), 117–123, 127–130, 133, 137, *see also* reactors

control volume, 64–69, 103, 118, 126
coordinated compound, 58–61
covalent bonding, 9, 12–13
critical depth, 81–82, *see also* Froude number
critical flow, 82, *see also* Froude number
critical velocity, 82, *see also* Froude number

D

DeBye-Hückel, 30, *see also* activity coefficient
diffusion, 92–93, 134, *see also* Fick's First Law
dispersion, 92–93, 134
drag coefficient, 91, 92, *see also* drag force
drag force, 74, 90–92
drought, 6, 21–23, 74–76

E

ecosystem function, 5–7, 113
ecosystem services, 5, 110–112
ecosystem structure, 3–5, 112–113
electrostatic forces, 12–13, 29, 102
ephemeral stream, 7, 97, 104, *see also* constancy of flow
equilibrium, 42–43, 49–60
equilibrium reaction constant, 42
evaporation, 17, 19–20, 67–72, 104, *see also* hydrologic cycle
evapotranspiration, 19–20, 67–68, *see also* hydrologic cycle

F

Fick's First Law, 92–93, *see also* diffusion
Froude number, 81–84, 85, 107–108

G

geomorphology, 6, 23, 97–117, 204
glide, 104, *see also* habitat
gravitational force, 73, 77, 90–91
gravitational potential energy, 73–74
Great Lakes, 21
groundwater, 6, 17, 20–22, 65, 67–68, 97, 103–104, 112, *see also* hydrologic cycle

H

habitat, 23–26, 34, 104–108, 112–113
habitat fragmentation, 7, 26
hardness, 14, 32–34